Hands-on Azure Cognitive Services

Applying AI and Machine Learning for Richer Applications

Ed Price
Adnan Masood
Gaurav Aroraa

apress®

Hands-on Azure Cognitive Services: Applying AI and Machine Learning for Richer Applications

Ed Price
Redmond, WA, USA

Adnan Masood
Temple Terrace, FL, USA

Gaurav Aroraa
Noida, India

ISBN-13 (pbk): 978-1-4842-7248-0
https://doi.org/10.1007/978-1-4842-7249-7

ISBN-13 (electronic): 978-1-4842-7249-7

Managing Director, Apress Media LLC: Welmoed Spahr
Acquisitions Editor: Smriti Srivastava
Development Editor: Laura Berendson
Coordinating Editor: Shrikant Vishwakarma

Cover designed by eStudioCalamar

Cover image designed by Pexels

Distributed to the book trade worldwide by Springer Science+Business Media LLC, 1 New York Plaza, Suite 4600, New York, NY 10004. Phone 1-800-SPRINGER, fax (201) 348-4505, e-mail orders-ny@springer-sbm.com, or visit www.springeronline.com. Apress Media, LLC is a California LLC and the sole member (owner) is Springer Science + Business Media Finance Inc (SSBM Finance Inc). SSBM Finance Inc is a **Delaware** corporation.

For information on translations, please e-mail booktranslations@springernature.com; for reprint, paperback, or audio rights, please e-mail bookpermissions@springernature.com, or visit http://www.apress.com/rights-permissions.

Apress titles may be purchased in bulk for academic, corporate, or promotional use. eBook versions and licenses are also available for most titles. For more information, reference our Print and eBook Bulk Sales web page at http://www.apress.com/bulk-sales.

Any source code or other supplementary material referenced by the author in this book is available to readers on GitHub via the book's product page, located at www.apress.com/978-1-4842-7248-0. For more detailed information, please visit http://www.apress.com/source-code.

Printed on acid-free paper

Thank you to Adnan and Gaurav for bringing their expertise, passion, and discipline to this book. I dedicate my portion of this book (it's probably whatever your favorite part is) to my oldest daughter, Eve. I wrote it when her teenage angstiness hit an all-time high, and she still let me hug her and hang out with her.

—Ed

Hi Mom and Dad!
Waves

—Adnan

My sincere thanks to Ed and Adnan. Without them, this book wouldn't be complete. The year was full of surprises and a lot of struggle throughout the globe. I dedicate this write-up to all the warriors of this era who came forward and helped to save humanity from this pandemic. And to my angel (daughter) Aarchi Arora whose smile always encourages and strengthens me to make me live for this project.

—Gaurav

Table of Contents

About the Authors

Ed Price is Senior Program Manager in Engineering at Microsoft, with an MBA degree in technology management. Previously, he led Azure Global's efforts to publish key architectural guidance, ran Microsoft customer feedback programs for Azure Development and Data Services, and was a technical writer at Microsoft for six years, helping lead TechNet Wiki. Ed now leads Microsoft's efforts to publish reference architectures on the Azure Architecture Center (including a strong focus on AI architectures). He is an instructor at Bellevue College, where he teaches design and computer science. At Microsoft, he also helps lead volunteer efforts to teach thousands of students how to code each year, focusing on girls and minorities. Ed is a co-author of six books, including *Azure Cloud Native Architecture Mapbook, Cloud Debugging and Profiling in Microsoft Azure* (Apress), and *Learn to Program with Small Basic*.

Adnan Masood, PhD, is an artificial intelligence and machine learning researcher, software engineer, Microsoft Regional Director, and Microsoft MVP for Artificial Intelligence. An international speaker and thought leader, Adnan currently works at UST as Chief AI Architect and collaborates with Stanford Artificial Intelligence Lab and MIT AI Lab on building enterprise solutions. Adnan has authored four books, including *Automated Machine Learning* and *Cognitive Computing Recipes* (Apress).

Gaurav Aroraa is Chief Technology Officer at SCL, with a doctorate in Computer Science. Gaurav is a Microsoft MVP award recipient. He is a lifetime member of the Computer Society of India (CSI), an advisory member, and senior mentor at IndiaMentor, certified as a Scrum trainer and coach, ITIL-F certified, and PRINCE-F and PRINCE-P certified. Gaurav is an open source developer and a contributor to the Microsoft TechNet community. He has authored ten books, including *Cloud Debugging and Profiling in Microsoft Azure* (Apress).

About the Technical Reviewer

 **Rohit Mungi** is an Azure Service Engineer with Microsoft from Hyderabad, India, with more than 13 years of experience in various roles in development, support, and DevOps. His passion for cloud technologies helped him understand and bridge the gap between on-prem and cloud migrations for various customers. His regular day work involves helping customers onboard new cloud technologies in the Data and AI (artificial intelligence) space, particularly AI and machine learning offerings of Azure. He is very active on Microsoft Q&A forums and ready to learn and share his knowledge on these topics. He has also been recognized as a community ninja on Microsoft's tech community. Follow him on LinkedIn at @rohit-mungi-0953678b.

Acknowledgments

To Rohit, our technical editor, for being so thorough, meticulous, detailed, careful, conscientious, diligent, scrupulous – we are running out of synonyms, but seriously, you rock.

To Shrikant for his extreme patience and regular follow-ups; thank you for putting up with us. Ghosting you is virtually impossible; we tried!

To Smriti for believing in us and this project!

Introduction

In about 400 BC, the Greek tale of Talos was told (and vividly illustrated). It was later chronicled in the epic poem, Argonautica, and its namesake is a giant automaton made of bronze, by Hephaestus, as requested by his father Zeus. This first robot (mythically speaking) sported wings, protected Europa (a woman) in Crete (an island), threw rocks at ships, and met his demise when Medea's spirits drove him crazy enough to pull out his own nail (thus, disassemble himself). In the 300s (BC), Aristotle brought us his syllogistic logic, a deductive reasoning system. Hero of Alexandria applied mechanics, basic robotics, and wrote *Automata* in the first century AD. AI in robots continued in myths with Pandora, who was also created by Hephaestus (you know, it's the story about the box), Pygmalian's Galatea (thanks to Venus, this is a statue of a woman who came alive to bear a child; don't ask), and then Jewish lore brought us Golems (in the early 1200s, AD), who are intelligent clay automatons that cannot speak.

Likely, in 1495, Leonardo da Vinci expanded on the many clockmakers' mechanical novelties to construct an automaton knight, and in 1515, we believe he continued his efforts to create a walking lion (presented to Francis I, the new King of France, and it's said to have been able to open its chest to present the gift of lilies). In 1642, Pascal invented a mechanical calculator. Swift's *Gulliver's Travels* (1726) seemed to have described an engine that's an AI computer and that forces knowledge on the poor Lilliputians. Punched cards came out in 1801, Mary Shelley scared us all with her intelligent *Frankenstein* in 1818, and in 1884, IBM got its start when Herman Hollerith invented a punched card tabulating machine to conduct the 1890 US census (that's the simplified explanation). At the 1939 World's Fair (in New York), Westinghouse unveiled Elektro, the mechanical man (he smoked), and Sparko, his mechanical dog.

Isaac Asimov wrote about robotics in 1950. Walt Disney (after having his Imagineers hack European wind-up novelties) developed a Dancing Man in 1961, he unveiled his tiki birds (which are in the tiki room) in 1963, he also recreated Abe Lincoln in 1963 for the 1964–1965 World's Fair, and he put the first animatronics in film with a bird in 1964's *Mary Poppins*. Ever since then, as computers evolved, we've been focused on building increasingly improved artificial intelligence in software and robotics (and we've been really hoping that our intelligent creations never take over and destroy us, like in *Terminator*, *Matrix*, or even *WALL-E*).

We intended this book to act as an introduction to Azure Cognitive Services, but also for it to become your guide to implementing Cognitive Services in real-world application scenarios. That means that we take you on a tour of everything that Cognitive Services has to offer, and, as you can imagine, that's quite a journey.

Azure Cognitive Services are broken down into services and APIs that explore decision, language, speech, and vision. With these APIs and technologies, you can create the future software that powers our computers and robots for centuries to come! (Or you can just make a silly website.) In this book, we take you on a journey to build an application, one step at a time. Please join us. After all, you're now walking in the footsteps of Aristotle, Hero, Hephaestus, Rabbi Loew (chief Golem mischief maker), Leonardo (not the turtle), Pascal, Gulliver, Mary Shelley, Hollerith, Elektro, Asimov, Disney, Bill Gates, Steve Jobs, and countless others. It's time to make a creation that's all your own.

CHAPTER 1

The Power of Cognitive Services

The terms **artificial intelligence (AI)** and **machine learning (ML)** are becoming more popular every day. Microsoft Azure Cognitive Services provides an opportunity to work with the top cutting-edge AI and ML technologies. To work with these technologies, we require some framework.

The aim of this first chapter is to set up the values, reasons, and impacts that you can achieve through Azure Cognitive Services. The chapter provides an overview of the features and capabilities. In the upcoming sections, you will understand how Azure Cognitive Services is helpful and how it makes it easy for you to work with AI and ML.

We also introduce you to our case study and the structures that we'll use throughout the rest of the book.

In this chapter, we cover the following topics:

- Overview of Azure Cognitive Services
- Exploring the Cognitive Services APIs: Vision, Speech, Language, Web Search, and Decision
- Overview of machine learning
- Understanding the use cases
- The COVID-19 SmartApp scenario

Overview of Azure Cognitive Services

Microsoft Azure Cognitive Services provides you with the ability to develop smart applications. You can build these smart applications with the help of APIs, SDKs (software development kits), services, and so on.

© Ed Price, Adnan Masood, and Gaurav Aroraa 2021
E. Price et al., *Hands-on Azure Cognitive Services*, https://doi.org/10.1007/978-1-4842-7249-7_1

Microsoft Azure Cognitive Services is a set of APIs, SDKs, and services that facilitate developers to create smart applications (without the prior knowledge of AI or ML).

Azure Cognitive Services provides everything that developers need in order to work on AI solutions, without the knowledge of data science. A developer can create a smart application that can converse, understand, or train itself.

Why Azure Cognitive Services

Azure Cognitive Services is backed by world-class model deployment technologies, and it is built by top experts in the area. There are a lot of plans and offers that use the pay-as-you-go model. You no longer have to invest in the development and infrastructure that you may need in order to build and host your models. Cognitive Services provides all this for you.

The following list shows the advantages you gain when you use Azure Cognitive Services:

- You don't need to build your own custom machine learning model.

- You gain a required AI service for your app. Azure Cognitive Services, as a **Platform as a Service (PaaS)**, can offer these required features.

  *You can build upon a **Platform as a Service (PaaS)** without being concerned about the infrastructure used to support the service.*

- You can invest your development time in the core app and release a stronger product.

Note   You should not use Azure Cognitive Services if it doesn't meet your requirements. For example, your data might have regulatory requirements that stop you from using an external service, like Azure. Or your organization might have a long-term commitment toward developing its own data science practices and product.

In the next section, we will discuss the Cognitive Services APIs in more detail.

Exploring the Cognitive Services APIs: Vision, Speech, Language, Web Search, and Decision

In the preceding section, we discussed Cognitive Services and the advantages that it provides. In this section, we will explore the APIs that are available to help developers.

Figure 1-1 provides a pictorial overview of these APIs.

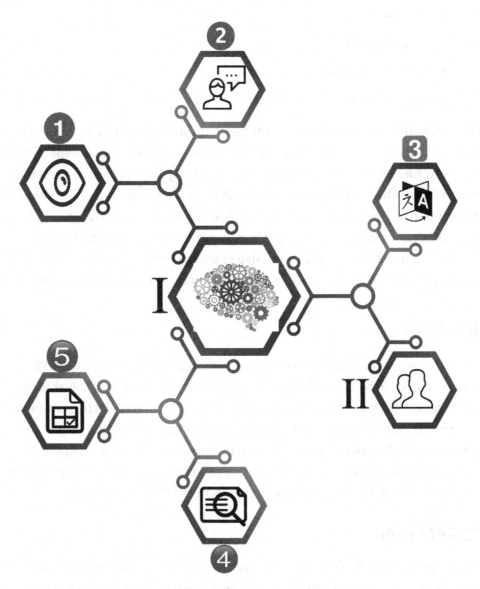

Figure 1-1. *Pictorial overview of Cognitive Services APIs*

Figure 1-1 displays the following elements of the Azure Cognitive Services APIs:

I – Represents all the Azure Cognitive Services APIs

II – Represents the developer who consumes these APIs to build smart apps

1 – Vision APIs

2 – Speech APIs

3 – Language APIs

4 – Web Search APIs

5 – Decision APIs

The Vision APIs provide insights on images, handwriting, and videos. The Speech APIs analyze and convert audio voices. The Language APIs offer you text analysis, they can make text easier to read, they help you create intelligent chat features, and they can translate text. The Bing Web Search APIs allow you to search and pull content from the entire Internet, leveraging pages, text, images, videos, news, and more. Finally, the Decision APIs help your app make intelligent decisions to moderate and personalize content for your users, and they help you detect anomalies in your data.

In the upcoming sections, we will briefly introduce you to each of these sets of APIs.

Vision APIs

First, let's explore the Vision APIs from Azure Cognitive Services. Use these APIs whenever you need to work with images or videos, to understand or analyze their contents. These APIs help you to get information like a facial analysis (determining age, gender, and more), feelings (e.g., through facial expressions), and more visual contents. Furthermore, with the help of these APIs, you can read the text from images, and thumbnails can be easily generated from images and/or videos. The Cognitive Services Vision APIs are divided into the following APIs, as detailed in the following.

Computer Vision

The Computer Vision API allows the developer to analyze an image and its contents. In the previous section, we discussed that with the help of Vision APIs, you can understand and collect the image contents. You can decide what content and information to retrieve from an image, based on your requirements. For example, a business might need to access the images in order to help make sure kids using their web app will avoid viewing adult content.

This API can also read printed text, hand-written text via optical character recognition (OCR).

Note   The current version of the Computer Vision API is v3.0, at the time of writing this book.

From a development perspective, you can either use RESTful (representational state transfer) APIs or you can build applications using an SDK. We will cover the development instructions and details in Chapter 3.

Custom Vision

The Cognitive Services APIs for Custom Vision provide a way to customize images with various customizations. You can customize the images with labels, and you can assess and improve these images based on the customization classifiers. The Custom Vision APIs use machine learning algorithms and apply labels to assess and improve the images. Furthermore, it is divided into two parts:

1. **Image classification** – Applies the label to an image.

2. **Object detection** – Applies the label, and it returns the coordinates from the image where the label is located.

Face

This API helps you detect and analyze human faces from an image. The algorithm can detect and analyze the data.

This service provides the following features:

- **Face detection** – Detects a human face and provides the coordinates of where the face is located in the image. Based on the algorithm, you can also get various properties of face detection, such as gender, the head pose, emotions, age, and so on. Figure 1-2 shows the face detection of a human (the author, Gaurav Aroraa).

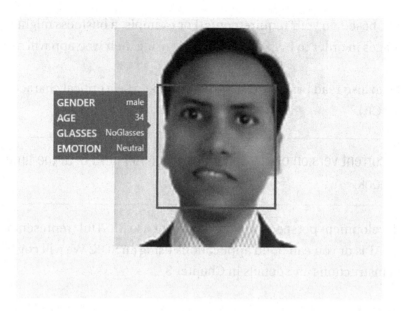

Figure 1-2. *Human face detection using the Face API*

- **Face verification** – Verifies two similar faces from images of one human face to compare them, in order to find out whether it belongs to the same human. Figure 1-3 shows two faces of the same human.

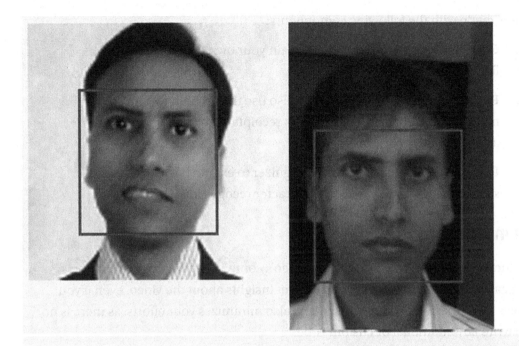

Verification result: The two faces belong to the same person. **Confidence is 0.82808.**

Figure 1-3. *Face verification*

- **Face grouping** – Groups the similar faces from an available database or set of faces.

Note   During your development cycle, the Face API and its data must meet the requirements of the privacy policies. You can refer to the Microsoft policies on customer data here: `https://azure.microsoft.com/en-us/support/legal/cognitive-services-compliance-and-privacy/`.

Form Recognizer

Form Recognizer extracts the data in key/value pairs and extracts the table data from a form-type document.

This is made with the following components:

- **Custom models** – Enables you to train your own data by providing five-form samples.

- **Prebuilt receipt model** – You can also use the prebuilt receipt model. Currently, only English sales receipts from the Unites States are available.

- **Layout API** – It enables Form Recognizer to extract the text and table structure data, by using optical character recognition (OCR).

Video Indexer

Video Indexer provides a way to analyze a video's contents, by using three channels: voice, vocal, and visual. In this way, you will get insights about the video, even if you don't have any expertise on video analysis. It also minimizes your efforts, as there is no need to write any additional or custom code.

Video Indexer provides us a way to easily analyze our videos, and it covers the following categories:

- Content creation

- Content moderation

- Deep search

- Accessibility

- Recommendations

- Monetization

We will cover video analysis more thoroughly in Chapter 3.

Speech APIs

Speech APIs provide you a way to make your application smarter. Thus, your application can now listen and speak. These APIs filter out the noise (words and sounds that you don't want to analyze), detect speakers, and then perform your assigned actions.

Speech Service

Microsoft introduced the Speech service to replace the Bing Speech API and Translator Speech. These are the services that provide an extraordinary effect to your application, in such a way that your application can hear users and speak/interact with your users.

Note   You can also customize Speech services by using frameworks. For speech to text, refer to `https://aka.ms/CustomSpeech`. For text to speech, refer to `https://aka.ms/CustomVoice`.

The Speech service enables the following scenarios:

- Speech to text
- Text to speech
- Speech translation
- Voice assistants

With the help of different frameworks, you can also customize your Speech experience.

Speaker Recognition (Preview)

Speaker Recognition is in Preview, at the time of writing this book. This service enables you to recognize the speakers; you can determine who is talking. With the help of this service, your application can also verify that the person that is speaking is who they claim to be. So, it is now much easier for your application to identify unknown speakers from a group of potential speakers.

It can be divided into these two parts:

- Speaker verification
- Speaker identification

We will cover voice recognition in detail in Chapter 5.

Language APIs

With the help of prebuilt scripts, the Language APIs enable your application to process the natural language. Also, they provide you the ability to learn how to recognize what users want. This would add more capabilities to your application, like textual and linguistic analysis.

Immersive Reader

Immersive Reader is a very intelligent service that builds a tool to help every reader, especially people affected with dyslexia.

Note   Dyslexia affects that part of the brain that processes language. People with dyslexia have difficulty reading, and they can find it very challenging to identify the sound in written speech.

Immersive Reader is designed to make it easier for everyone to read. It provides the following features:

- Reads textual content out loud

- Highlights the adjectives, verbs, nouns, and adverbs

- Graphically represents commonly used words

- Helps you understand the content in your own translated language

Language Understanding (LUIS)

Think of a scenario where you need to make your application smart enough, so that it can understand user input (such as speech, text, and so on). The Speech service makes your application smart enough to listen and speak with the user. But your application might need to be smart enough to answer a question that your user asks it, such as, *"What is my health status?"* Even after implementing the Speech APIs, your application will not be ready to understand commands like that. To achieve such a complicated requirement, we have the Language Understanding (LUIS) services. (LUIS stands for Language Understanding Intelligent Service.) With the help of LUIS, you can build an application that interacts with users and pulls the relevant information out of the conversations. For a question like, *"What is my health status?",* your application can assess the stored data, and it then provides the status of the user's health. Or it asks a few questions, and based on the user's answer, it would then provide the user's health status.

You can work with the following two types of models:

- Prebuilt model
- Custom model

Learn more about LUIS in Chapter 4.

QnA Maker

QnA is very relevant, when you have an FAQ and want to make it interactive. This means that you have a predefined set of QnA (questions and answers). QnA is mostly used in chat-based applications, where the user enters queries, and then your application answers the question. You can try using Microsoft's `www.qnamaker.ai/` to enable your experience with QnA Maker.

Text Analytics

With the help of the Text Analytics service, you can build an application that analyzes the raw text and then gives you the result. It includes the following functions:

- Sentiment analysis
- Key phrase extraction
- Language identification
- Name identifications

Translator

Translator enables text-to-text translation, and it provides a way to build translation into your application. With the help of Translator, you can add multilingual capabilities to your application. Currently, more than 60 languages are supported. If you want to translate a spoken speech, you will need to use the Speech service.

Web Search APIs

The Web Search APIs enable you to build more intelligent applications, and they give you the power of Bing Search. They allow you to access data from billions of web pages, images, and news articles (and more), in order to build your search results.

Bing Search APIs

Bing Search facilitates your application by providing the ability to do a web search. You can imagine that with the implementation of Bing Search APIs, you now have a wide range of web pages with which to build out your search results. The code implementation is very easy as well (see Listing 1-1).

Listing 1-1. The sample code to implement Bing Web Search

```
//Sample code
public static async void WebResults(WebSearchClient client)
{
    try
    {
        var fetchedData = await client.Web.SearchAsync(query: "Tom
        Campbell's Hill Natural Park");
        Console.WriteLine("Looking for \"Tom Campbell's Hill Natural
        Park\"");

        // ...

    }
    catch (Exception ex)
    {
        Console.WriteLine("Exception during search. " + ex.Message);
    }
}
```

Bing Web Search

With the Bing Web Search API, you can suggest search terms while a user is typing, filter and restrict search results, remove unwanted characters from search results, localize search results by country, and analyze search data.

Bing Custom Search

The Bing Custom Search API allows you to customize the search suggestions, the image search experience, and the video search experience. You can share and collaborate on your custom search, and you can configure a unique UI for your app to display your search results.

Bing Image Search

The Bing Image Search API enables you to leverage Bing's image searching capabilities. You can suggest image search terms, filter and restrict image results, crop and resize images, display thumbnails of the images, and showcase trending images.

Bing Entity Search

The Bing Entity Search API gets you search results of entities and locations, such as restaurants, hotels, or stores. You can provide real-time search suggestions, entity disambiguation (providing multiple search results), and return information on businesses and other entities.

Bing News Search

The Bing News Search API lets you search for relevant news articles. You can suggest search terms, return news articles, showcase the day's top news articles and headlines, and filter news results by category.

Bing Video Search

The Bing Video Search API returns high-quality videos. You can suggest search terms, filter and restrict the results of the video query, create thumbnail previews of the videos, showcase trending videos, and access video insights.

Bing Visual Search

The Bing Visual Search API provides metrics and insights for an individual image, whether one that is uploaded or shared via a URL. You can identify similar images and products, identify shopping sources for the image, access related searches made by others (based on the content of your image), find the web pages that display the image, find recipes for a dish featured in your image, and use the image to automatically gather information about an entity (such as information about an actor in an image or directions to a location in the image).

Bing Autosuggest

The Bing Autosuggest API can improve your search experience by returning a list of suggested queries, based on a partial query.

Bing Spell Check

The Bing Spell Check API performs grammar and spell checking on any text. You can check for slang or informal language, differentiate between similar words, and track new brands, titles, and popular expressions.

Bing Local Business Search

The Bing Local Business Search API helps you find information about a business. You can find the nearest restaurant to you (such as a specific chain or a type of food), locate and map out specific businesses and places, limit the distance parameters of your search, and filter the business results by category.

Decision APIs

Decision APIs help you create a rich user experience that is personalized, with content moderation. They help you perform efficient decision making.

Anomaly Detector (Preview)

Anomaly Detector is in Preview, at the time of writing this book. It helps you detect abnormalities in the data, in order to remove them. When you use this service, you don't need to think about what model is fit for your business. The Anomaly Detector service automatically determines the best-fit model of your data.

Content Moderator

Content Moderator helps you identify the content that your business does not allow or content that does not fit your business. It checks text, image, and video contents as well. For example, you can remove profanity and undesirable text, you can identify videos that contain profanity, you can moderate adult and racy video and image content, you can check text for personally identifiable information (PII), and you can detect other offensive or unwanted images.

Personalizer

Personalizer helps you give a personal experience to your user individually. It helps you show different and personalized content, to individual users. The content can be within text, images, URLs, or emails.

Overview of Machine Learning

There are a lot of possible definitions of what machine learning is. According to all the available definitions, we can conclude the following.

Machine learning is a subset AI that provides you with a way to study computer algorithms. In this process, after studying the data, computer programs can forecast future events.

Our learning process is simple and continuous. The process happens anytime and anywhere, like while you're walking, making a coffee, or purchasing items. Every learning has its own scope and boundaries. For example, while you are walking, your smartwatch (a machine fitted with a program) collects your data (such as step counts, miles, heartbeats, and your blood pressure). For another example, while you visit the department store for purchases, technology systems (like loyalty cards and supporting software) learn about your purchase behavior, based on the items that you've purchased in the past.

With this information, machines can predict the future. For example, the machine that learns purchase behavior can predict which items would be ordered by their customers. For another example, it helps the store owners understand how much a customer consumes a particular item.

Note   In summary, machine learning is a study that helps us understand future behavior. We achieve this by studying data in various aspects, such as analyzing text for emotional sentiment or analyzing images to recognize objects or faces.

There are several phases in the machine learning process. These phases are iterative and require repetition, based on the requirement and/or training. Here are the most common phases:

- **Data collection phase** – In this phase, the data is being prepared or collected. You determine what data to collect, based on the need of training or the component that is required for training.

- **Test and develop training models** – This is the phase where the learning is being tested and where the training models are developed. For example, to build a new training model for a department store, we need to analyze the grocery items. You would gather and access the data by using algorithms.

- **Deploy and manage trained models** – This is the phase where you can see the actual output. You can also call this phase *machine learning in action.*

Microsoft made this easy for developers, with the help of various plans. You can follow the preceding phases and manage your machine learning models, using the available Microsoft technologies.

The following various Microsoft technologies are available, to work with ML:

- **Cloud based** – Microsoft Azure provides a lot of options, where you can use cloud services and work with machine learning.

  - Cognitive Services APIs (such as Vision, Speech, Language, and Web Search) provide sophisticated pretrained ML models.

- **On-premises** – Similar to cloud-based options, you have on-premises options where local (on-premises) servers can also be run in virtual machines. The most commonly used services are as follows:

  - SQL Server Machine Learning services allows you to run Python and R scripts with relational data, for predictive analytics and machine learning.

  - Microsoft Machine Learning Server runs in-database analytics in SQL Server and Teradata.

- **Tools**

  - Azure Data Science Virtual Machines (DSVMs) are preinstalled, configured, and tested with various ML and AI tools, to optimize support of ML and AI workloads.

  - Azure Databricks is Apache Spark based, scales analytics, and unlocks insights from data and AI solutions.

  - ML.NET is an open source and cross-platform ML framework that supports Windows, Linux, and macOS. It is built for .NET developers, and it works seamlessly with ML libraries, such as TensorFlow, ONNX, and Infer.NET.

  - Windows ML is built into Windows 10 and Windows Server 2019. It is an API that deploys ML on Windows devices that are optimized for this purpose.

  - MMLSpark is an open source set of tools that provides seamless integration with Azure Spark and the like.

  - Additional frameworks include PyTorch, Keras, ONNX, and TensorFlow.

- **Azure Machine Learning** – This is a service, based on the Azure cloud. It provides a way to train, deploy, and manage machine learning models. (To get started with Azure Machine Learning, go to https://ml.azure.com.) The main advantage of this is that it is fully open source, and so you can take advantage of the following technologies:

- Azure notebooks is a free service for R developers who work with notebooks and bring their code and apps.

- Jupyter notebooks is an open source project for developers to grow and learn proactively.

- Azure Machine Learning can answer your extension questions for Visual Studio Code.

You can sign up at `https://aka.ms/AMLFree` *for 12 months of free Azure Machine Learning services.*

Understanding the Use Cases

Machine learning (ML) is getting more popular each day. More and more people are using it in their business requirements. The following industries prolifically use ML, to forecast their business needs, as well as artificial intelligence (AI) in general:

- **Finance and banking** – ML is being used to understand the needs of future borrowers for credit decisions, to personalize loan options, as well as for determining loan eligibility. In addition, credit scoring is being driven by AI. Risk management is completely dependent on AI, in order to manage a massive amount of structured and unstructured data (by analyzing the history of issues, failures, and successes). Banking and leasing companies immediately find significant improvements when using cloud-based AI services to determine risk, rather than using their previous data science processes. AI and ML are also used for fraud prevention, trading, personalized banking, process automation, digital assistants, and account transaction security.

- **Fraud detection** – Although the number of identity thefts are growing each year, AI and ML are being applied to countless data points in order to analyze and compare a consumer's historical record and data, in order to identify and stop fraudulent transactions. As this trend continues, financial fraud is being caught more often and much sooner than ever before.

- **Healthcare** – AI is being used for robotic surgery, virtual nursing assistants, and automating workflow and administrative activities. AI and ML are also used to make clinical judgments and to help diagnose patients. This includes detecting skin cancer, breast cancer, and an early detection and prediction of cancer in general, by examining medical records, habits, and genetic information. AI is also being used to identify cardiac arrests, based only on the tone of someone's voice and the background noise in phone calls. AI is also being used for image analysis (such as CT scans), which results in an increased reliability in accurate detection and diagnosis of various inflictions, and it is already helping reduce hospital wait times.

- **Retail, ecommerce, and advertising** – For the retail industry, ML is being used for supply chain planning, demand forecasting, customer intelligence, marketing campaigns and advertising, store operations, pricing, and product promotions. Stores are using loyalty cards in order to track who purchases what items. They predict their sales for every day of the calendar year, based on previous days in all the previous years, taking into account recent and annual trends and changes. The data is also used to staff appropriately, planning schedules and vacation hours around the demand, forecasting down to the time of day. From the very beginning, that data has also been used by online retailers and websites to advertise to you, based on your online purchases, searches, websites you visit, and content you view.

- **Education** – AI is being used now to create unique, individualized learning systems. This intelligent instructional design provides students with challenges that are constantly meeting them where they are at, taking them to the next level the moment that they are ready. AI is enabling students universal access to global schools, including translating languages for students in order to understand foreign instructors. AI is also helping create learning opportunities for students who might have visual or hearing impairments. In addition, AI and ML are helping automate school and classroom administrative tasks, and they help provide tutoring and support to students outside the class.

19

The COVID-19 SmartApp Scenario

While we planned this book, we're facing a Coronavirus disease (COVID-19) pandemic: www.who.int/emergencies/diseases/novel-coronavirus-2019. We decided to draft our examples in such a way so that we'd cover real-world use cases.

The COVID-19 SmartApp covers the following scenarios:

- Data collection

- Data analysis

- Vision analysis

We'll revisit the COVID-19 SmartApp throughout this book, offering samples and explanations, in order to provide you with an example of an end-to-end smart application.

Summary

The aim of this chapter was to set up the values, reasons, and impacts you can achieve through Microsoft Azure Cognitive Services. This discussion started with an overview of Azure Cognitive Services and why you would use Cognitive Services. We followed this with high-level overviews of the various APIs provided by Cognitive Services. These APIs minimize our efforts related to machine learning and artificial intelligence. This chapter also covered machine learning and the various offerings from Microsoft, including Azure Machine Learning.

During this discussion, this chapter acknowledged the various use cases related to AI and ML. Finally, we briefly introduced you to our COVID-19 SmartApp.

In the next chapter, we will continue with the discussion of Azure Cognitive Services. We will discuss how to get started with Azure Cognitive Services by exploring the Microsoft Azure portal.

CHAPTER 2

The Azure Portal for Cognitive Services

This chapter will explore how you can get started with Cognitive Services on the Azure portal, and it includes an exploration of the common features. Next, the chapter will take you inside the Azure Marketplace for Bot Service, Cognitive Services, and Machine Learning.

In this chapter, we will cover the following topics:

- Getting started with the Azure portal and Azure Cognitive Services

- Azure Marketplace – an overview of AI + Machine Learning

- Understanding software development kits (SDKs) – to get started with your favorite programing language

- Setting up your Visual Studio template

Getting Started with Azure Portal and Azure Cognitive Services

Microsoft Azure Cognitive Services provides the facility to develop smart applications. You can build these smart applications with the help of APIs, SDKs, services, and so on.

Microsoft Azure Cognitive Services are a set of APIs, SDKs, and services that facilitate developers to create smart applications, without the prior knowledge of AI or ML.

© Ed Price, Adnan Masood, and Gaurav Aroraa 2021
E. Price et al., *Hands-on Azure Cognitive Services*, https://doi.org/10.1007/978-1-4842-7249-7_2

Azure Cognitive Services provides everything that a developer (without the knowledge of data science) needs, in order to work on smart applications. For example, with the help of Cognitive Services, a developer can create a smart application that can converse, understand, or train itself.

In this section, we will walk through the Azure portal and explore the various options of Azure Cognitive Services.

This section is a walkthrough of the Azure portal. You can skip this section if you're already familiar with the Azure portal.

To get started with the Azure portal, you need a valid Azure account. If you do not have a valid Azure account, follow these points, to register a free Azure account:

- Go to `https://signup.azure.com`.

- To sign up a free account, you must have a valid phone number, a valid credit card, and a GitHub account or Microsoft account Id (previously called a Windows Live ID).

- Follow the instructions on the screen.

After completing your registration, your free account consists of the following benefits:

- Free services for 12 months

- With the free account, you can build a smart/intelligent application, with the help of these free services:

    - Computer Vision

    - Personalizer

    - Translator

    - Anomaly Detector

    - Form Recognizer

    - Face services

    - Language Understanding

    - QnA Maker

To view the complete list of free products and services, refer to this link:
`https://azure.microsoft.com/en-us/free/`.

After completing your registration, sign in to the Azure portal (`https://portal.azure.com`). After you sign in, you will land on the Azure portal home page, which should look like Figure 2-1.

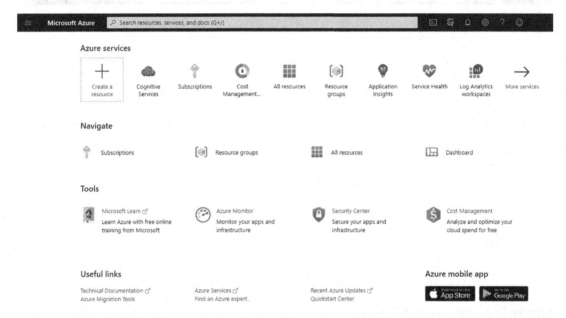

Figure 2-1. *The Azure portal*

You can see the various services available to you, as per your subscription. This is the starting point to explore the Azure products and services.

A complete walkthrough of the Azure portal is beyond the scope of this book. If you want to learn about working in the Azure portal in detail, please refer to the book, ***Cloud Debugging and Profiling in Microsoft Azure***, published by Apress.

Getting Started with Azure Cognitive Services

Let's explore Microsoft Azure Cognitive Services. Please note that we will cover each and every service with code examples in the coming chapters. To get started with Cognitive Services, from the Azure portal, search for "Cognitive Services" from the search text box, or click **All Services ➤ AI + machine learning ➤ Cognitive Services** (see Figure 2-2).

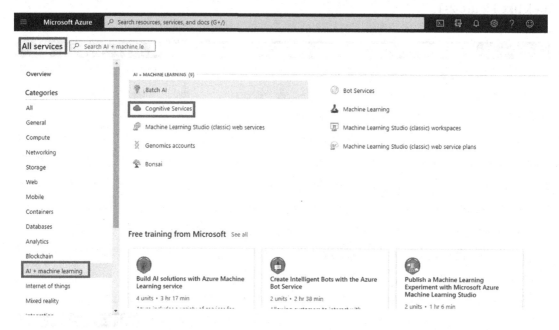

Figure 2-2. *Cognitive Services*

Next, you will see the Cognitive Services screen. From this screen, you can manage your existing service or add/create a new service. This is the starting point from where you will need to begin exploring the Azure Marketplace.

In the forthcoming section, we will discuss Cognitive Services in Azure Marketplace.

Azure Marketplace: An Overview of AI + Machine Learning

Azure Cognitive Services are backed by world-class model deployment technology, which is built by the top experts in the area. With the pay-as-you-go model, you can choose from a lot of different plans and offers. You can find a faster, easier, and often cheaper solution, instead of investing in the development and infrastructure that you will likely need (if you choose to develop and host your models for such common use cases).

Azure Marketplace is a one-stop shop to get all the services. You will be sent to Azure Marketplace, as soon as you click **Add** or **Create cognitive services** (see Figure 2-3).

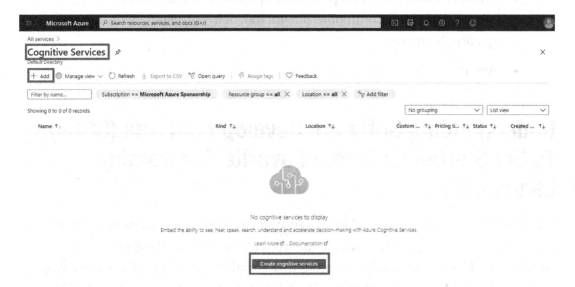

Figure 2-3. *Add or create cognitive service*

Azure Marketplace provides many categories of offerings, including these examples:

- **Bot Service** – Includes tools to create, test, deploy, and manage your intelligent bots.

- **Cognitive Services** – In Chapter 1, we provided an overview of these services.

- **ML service** – Azure Machine Learning builds intelligent models and repeatable workflows for you to deploy and manage.

- **Automated ML** – With AutoML, you can train and tune a model, by specifying your target metric.

- **Business or robotic process automation** – Allows you to create virtual workforces to help drive your business forward.

- **Data labeling** – Create, manage, and monitor your labeling projects, across a team.

- **Data preparation** – Ensure that the data uses the correct encoding and a consistent schema, for example.

- **Knowledge mining** – Quickly explore and learn from large amounts of data, to uncover important insights, relationships, and patterns.

- **ML operations** – MLOps applies DevOps to machine learning, including building continuous integration, delivery, deployment, and quality assurance.

- And more

Understanding Software Development Kits (SDKs): To Get Started with Your Favorite Programing Language

You would find it too complex and time consuming to write each and every piece of code from scratch. We can minimize our efforts and time with the help of software development kits (SDKs). You have various SDKs available that you can use to develop an application with Cognitive Services. The following major languages are currently supported by the SDKs (as of when we wrote this book):

- C#

- Go (often referred to as Golang, or Go Language)

- Java

- JavaScript

- Python

- R

We will use C# for all the examples in this book.

Setting Up Your Visual Studio Template

At this point, you have a basic understanding of the Azure portal and the various services provided in Azure Marketplace. Now it's time to start building a small app. In this section, you will start by setting up your environment, using Visual Studio, and you'll follow this approach throughout the book.

We are creating this template so that while we create more small applications to describe the topic in the coming chapters, we do not need to recreate this same application more than once.

We use Visual Studio 2019 Community Edition to build the examples.

Before you get started with the template setup, make sure you have **ASP.NET and web development** and the **.NET desktop development** workloads installed in your Visual Studio 2019 environment (see Figure 2-4).

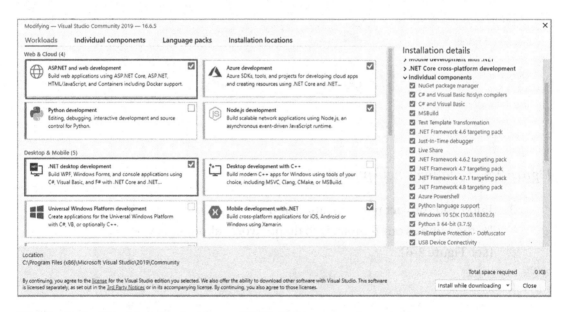

Figure 2-4. *Workload selection – .NET desktop development*

Let's create a Visual Studio template:

1. Open Visual Studio.

2. Click **File**, **Project**, **New Project**, and then search for and select **ASP.NET Core Web Application** (see Figure 2-5). Click **Next**.

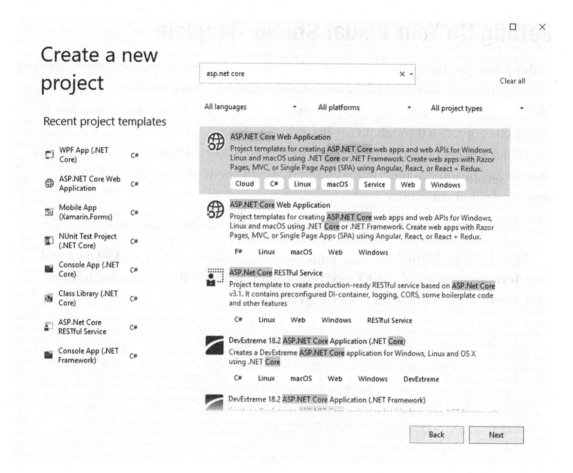

Figure 2-5. *Creating a new project*

3. Enter your project name location and so on. These values will not
   be the same as our template, so do not worry about these values
   (see Figure 2-6).

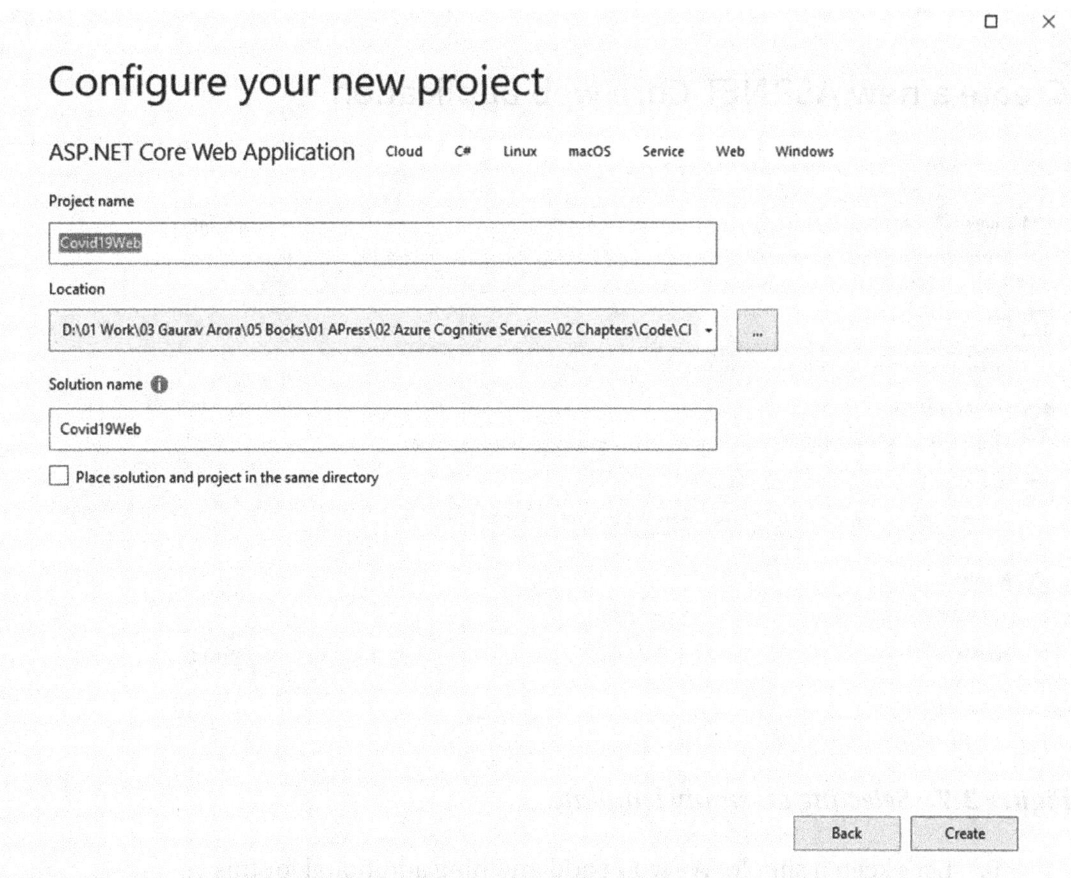

□   ✕

Configure your new project

ASP.NET Core Web Application   Cloud   C#   Linux   macOS   Service   Web   Windows

Project name

Covid19Web

Location

D:\01 Work\03 Gaurav Arora\05 Books\01 APress\02 Azure Cognitive Services\02 Chapters\Code\CI ▾   ...

Solution name ⓘ

Covid19Web

☐ Place solution and project in the same directory

Back   Create

Figure 2-6. *Configuring your new project*

4. Select the **Empty** project (in the left pane). Make sure you select
    .NET Core and **ASP.NET Core 3.1** from the drop-down lists, at the
    top of the page (see Figure 2-7). Click **Create** (in the lower right
    corner).

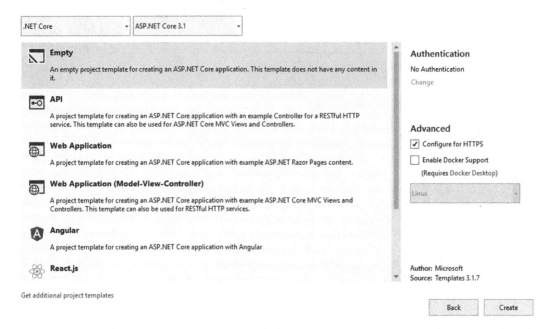

Figure 2-7. Selecting an empty template

5. Let's keep it simple. We won't add anything additional for this
    template. Click **Project** and then **Export Template**. From the
    Choose Template Type screen (in the Export Template Wizard),
    select the **Project template** option, and then click **Next** (see
    Figure 2-8).

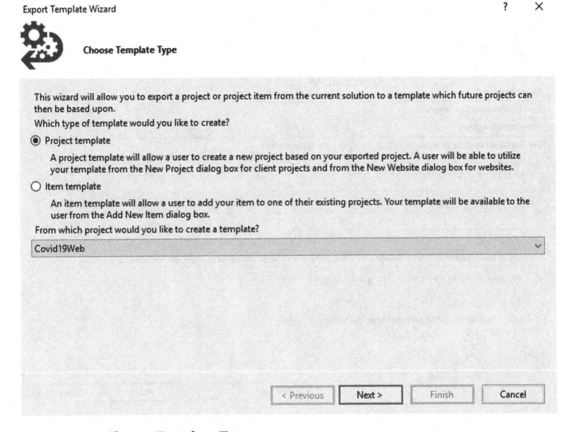

Figure 2-8. *Choose Template Type*

6. From the next screen of the Export Template Wizard, provide the
    Template name and **Template description**. You can also provide
    the **Icon Image**. Make sure both of the check boxes remain
    checked, and then click **Finish** (see Figure 2-9).

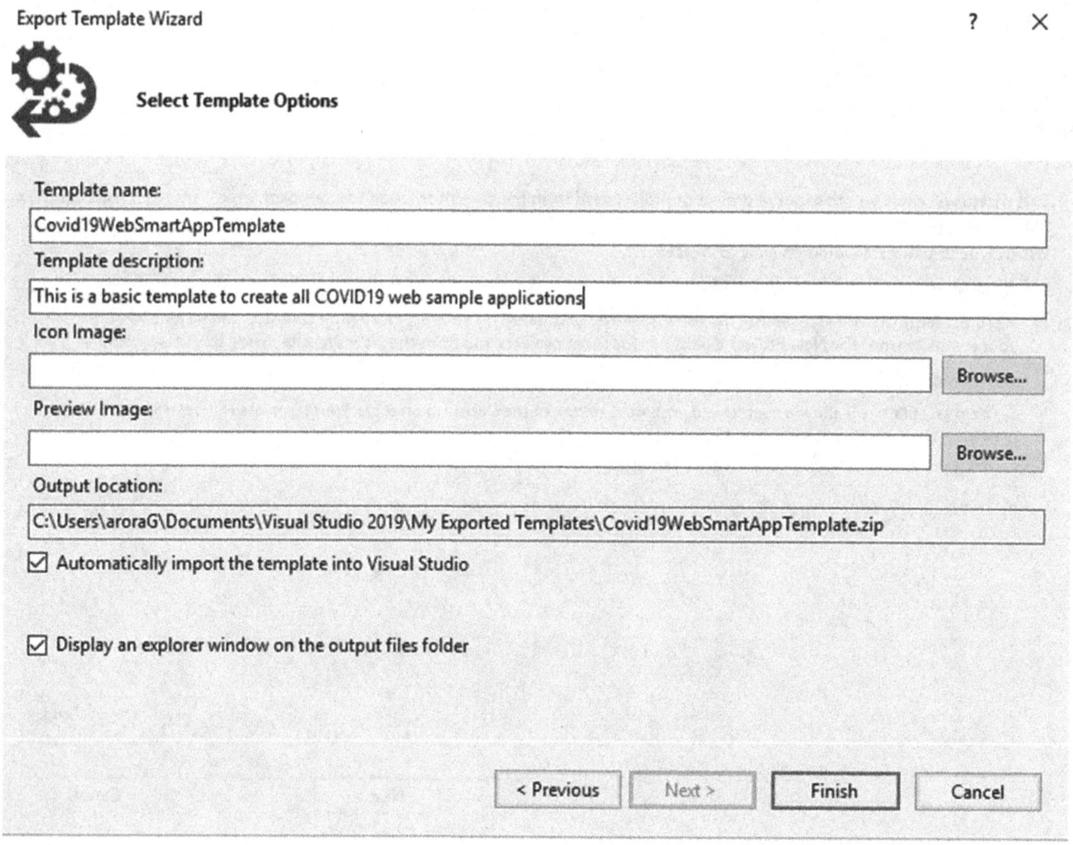

Figure 2-9. *Select Template Options*

Your exported template should be generated into your "My Exported Templates" folder in the "Visual Studio 2019" folder.

We have also created a template for a desktop application, by using WPF. This template (Covid19WpfApp.zip) is located in the GitHub repository https://github.com/Apress/hands-on-azure-cognitive-services/ blob/main/Chapter%2002/Templates/Covid19WpfApp.zip.

We're now done with the preparation of the template. This is a blank Asp.Net Core 3.1 template that we will use with our upcoming code examples. You can also extract the template zip file and open the MyTemplate.vstemplate file in any text editor. (We opened it in Notepad++.) The zip file contains the following files, as shown in Figure 2-10.

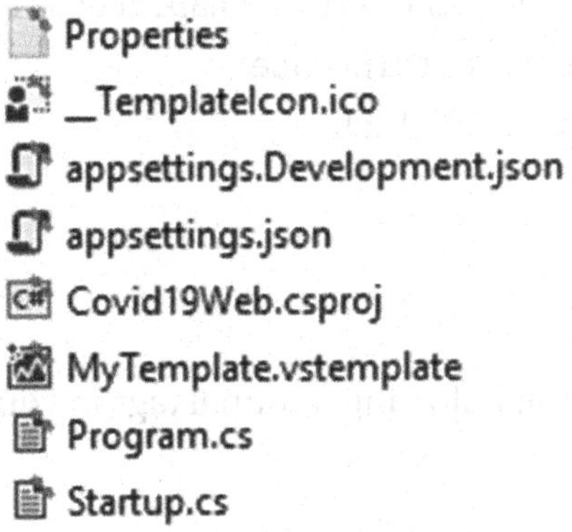

- Properties
- _TemplateIcon.ico
- appsettings.Development.json
- appsettings.json
- Covid19Web.csproj
- MyTemplate.vstemplate
- Program.cs
- Startup.cs

Figure 2-10. *Template files*

The MyTemplate.vstemplate file consists of the following code (see Listing 2-1).

Listing 2-1. MyTemplate configuration

```
<VSTemplate Version="3.0.0" xmlns="http://schemas.microsoft.com/developer/
vstemplate/2005" Type="Project">
  <TemplateData>
    <Name>Covid19 Web Smart App Template</Name>
    <Description>This is a basic template to create all COVID19 web sample
    applications</Description>
    <ProjectType>CSharp</ProjectType>
    <ProjectSubType>
    </ProjectSubType>
      <LanguageTag>C#</LanguageTag>
      <ProjectTypeTag>Web</ProjectTypeTag>
      <ProjectTypeTag>Covid19</ProjectTypeTag>
    <SortOrder>1000</SortOrder>
    <CreateNewFolder>true</CreateNewFolder>
    <DefaultName>Covid19WebSmartAppTemplate</DefaultName>
    <ProvideDefaultName>true</ProvideDefaultName>
    <LocationField>Enabled</LocationField>
```

```
    <EnableLocationBrowseButton>true</EnableLocationBrowseButton>
    <CreateInPlace>true</CreateInPlace>
    <Icon>__TemplateIcon.ico</Icon>
  </TemplateData>
  ...
</VSTemplate>
```

Make sure you update the following optional tags in your exported template file (from Listing 2-1):

- LanguageTag

- ProjectTypeTag(s)

You can see this is a simple XML file that defines the various properties of the template. The main properties are listed as follows. These properties help Visual Studio to group the project template on the Add New Project dialog:

- **Template Data** – This is to define the template name and its description. This name and description appear on the Add New Project dialog window.

- **ProjectType** – This contains the value of the project type, such as CSharp.

- **LanguageTag** – This is an important property that tells Visual Studio that the template is for a specific language. For example, our template is for C#.

- **ProjectTypeTag** – This contains the value of the project type, such as Web, Desktop, and so on. Our template has these ProjectType labels: Web and Covid19.

In this section, we created a Visual Studio template. This template will be helpful for us, while we create our code examples in the upcoming chapters.

Summary

The aim of this chapter was to explore the Azure portal and to get started with Cognitive Services. In this chapter, we discussed Azure Marketplace. We went through the various available SDKs. Finally, we created a Visual Studio template to start building an app with Cognitive Services.

In the next chapter, we will continue with the discussion of Azure Cognitive Services. We will discuss how to get started with Azure Cognitive Services, and we will further explore the Microsoft Azure portal.

CHAPTER 3

Vision – Identify and Analyze Images and Videos

The identification of images and videos is one of the core features needed to make a smart application. Azure Cognitive Services provides us with the Vision API, to work with images and videos.

The aim of this chapter is to get started with the Vision API and to create a smart application that helps you work with images and videos.

In this chapter, the following topics will be covered:

- Understanding the Vision API with Computer Vision

- Analyzing images

- Identifying a face

- Understanding the working behavior of Vision APIs for video analysis

- Summary of the Vision API

In coming sections, we will discuss, understand, and use the Computer Vision API.

Understanding the Vision APIs with Computer Vision

Computer Vision is a set of APIs that come under the category of artificial intelligence (AI). It helps you train computers (machines). The process of training a machine is to teach the computer how to understand the visual world. The act of capturing text data from the digital image (and capturing graphics data from videos) is one of the most powerful inventions in the world of Computer Vision.

© Ed Price, Adnan Masood, and Gaurav Aroraa 2021
E. Price et al., *Hands-on Azure Cognitive Services*, https://doi.org/10.1007/978-1-4842-7249-7_3

Artificial intelligence (AI) refers to the technique how machines can think like humans. This technique comes with programming, algorithms, and training, where machines are instructed or trained to think and behave like humans. The term AI was coined in 1956, when scientists explored topics like problem solving. It then grew in popularity in 1960 when Defense Advanced Research Projects Agency (DARPA) took interest in the work to train computers and mimic basic human reasoning. The research was so good that, in 2003, DARPA introduced intelligent personal assistants.

Deep learning is a type of machine learning that helps train the computers/machines, so that these machines perform tasks like humans do. These tasks could be speech recognition, image identification, object specifications, or the machines can take possible decisions and make predictions.

In short, Computer Vision comes under artificial intelligence (AI). With the help of training (called deep learning), the trained machines can easily identify the objects. A training model also helps these machines classify the identified objects. Note that the accuracy to identify and specify the objects will depend upon the training of the machines.

Optical character recognition (OCR) is a technique that helps you create editable or searchable data from visual, textual documents, such as scanned paper, images, PDF files, and or any other file format that contains textual or visual contents that can be extracted.

Computer Vision is not new. It has been here for more than six decades. It got more popular in the 1970s, when **Raymond Kurzweil** (known as Ray Kurzweil) commercialized the first optical character recognition tool, which was named **"omni-font OCR"**. This tool was able to process the printed text in almost any font. In the year 2000, the OCR was released as a cloud-based service.

Now, we can say that Computer Vision is mature (but keep in mind it's still growing and learning itself). It is one of the best gifts for technologists and scientists, to take advantage of the advanced capabilities to analyze and identify the data. The inner workings of Computer Vision are rather simple and can be defined in the following steps:

1. **Collecting images** – These days, we have a host of space, which can be 100 GB to 1 TB or more, and we have the technology to capture digital images. So, today it is very easy to capture and collect digital images, photos, and 3D objects.

2. **Image processing** – The first step to train the model is that we need data. This data is merely a collection of related images. Now, we have technology with more compute power. We can process thousands of pre-identified images.

3. **Image understanding** – This is the step where the Computer Vision has identified or classified the object.

The preceding steps are called thumb steps, which only describe the basic functionality of Computer Vision. These steps can further be divided, based on decisions made. Computer Vision can also be divided into the following steps:

- Image partitioning or separating an image into multiple regions, subregions, or pieces of a large image, so that each piece can be examined separately.

- Using X and Y coordinate detection or the identification of objects in a single image. With the help of X and Y coordinates, it creates a bounding box, and then it's easy to identify everything inside the box.

- Considering facial recognition, which is an advanced type of intelligence that detects the object. It is so advanced that it not only recognizes that it sees a human face in an image, but it also identifies a specific individual (whose face it is).

- Pattern detection is another advanced functionality, which recognizes repeated shapes and colors in the images.

Microsoft Azure Cognitive Services   The Computer Vision API is a set of cloud-based services that provide advanced algorithms and that help developers analyze and process the images, to retrieve the information. In short, the Computer Vision APIs provide insights on images, handwriting, and videos.

Now, OCR supports API v3.2 for 73 different languages. You can find this language information at `https://docs.microsoft.com/en-us/azure/cognitive-services/computer-vision/language-support#optical-character-recognition-ocr`.

To get started with the Azure portal, you need a valid Azure account. If you do not have a valid Azure account, take the following steps to register a free Azure account:

- Go to `https://signup.azure.com`.

- To sign up for a free account, you are required to provide a valid phone number, a valid credit card, and a GitHub account or Microsoft account ID (previously known as the Windows Live ID).

- Follow the instructions on the screen.

After registration, your account will include 12 months of free services.

You can then sign in to Azure, at `https://portal.azure.com`. Once you sign in, you will land on the portal home page, which currently looks like Figure 3-1.

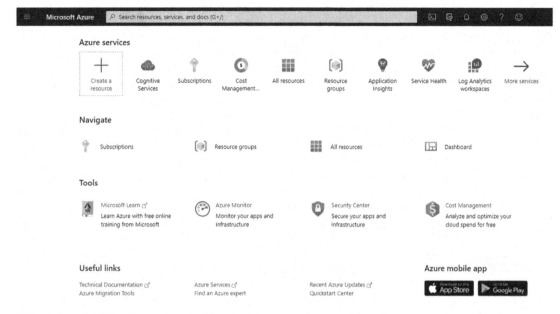

Figure 3-1. *The Azure portal*

A complete walkthrough of the Azure portal is beyond the scope of this book. If you want to learn the details of navigating and using the Azure portal, please refer to the book ***Cloud Debugging and Profiling in Microsoft Azure***, published by Apress.

Analyzing Images

In the previous section, we discussed how the Computer Vision APIs provide a way to identify and analyze object. Computer Vision does this to extract the data from images with the help of various algorithms. Computer Vision makes a developer's job easier, by providing hundreds of objects and parameters. In this section, we will learn more about the APIs, and we'll take it to the next level, with a code example.

First, go to the Azure portal, and search "Cognitive Services" from search text box. Alternatively, you can click **All services**, **AI + machine learning** (in the left pane), and **Cognitive Services** in the right pane. See Figure 3-2.

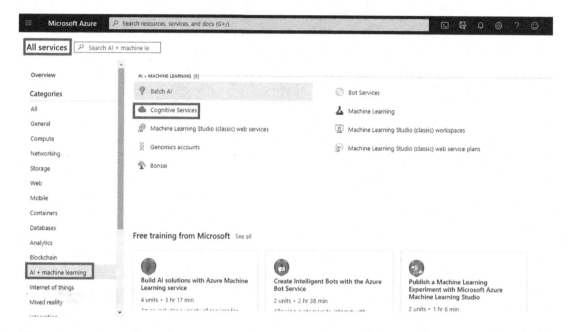

Figure 3-2. *Searching for Cognitive Services*

Next, you will see the Cognitive Services screen. From this screen, you can manage your existing service or add/create a new service. This is the starting point from where the journey of image analysis begins.

The Computer Vision Cognitive Service uses a model that is pretrained and helps developers analyze the images.

Start Diving for Computer Vision

To take advantage of the Computer Vision service, let's first create a resource from the Azure portal. Search "Computer Vision" from the search text box, at the top of the screen. From the search results, click **Computer Vision** under *Marketplace*, as shown in Figure 3-3.

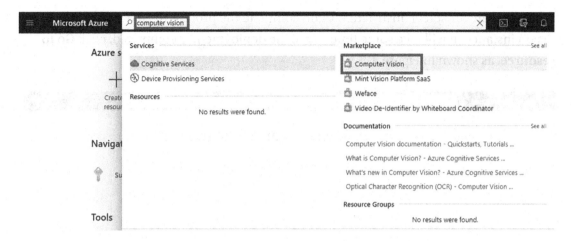

Figure 3-3. *Computer Vision*

On the Create Computer Vision page, you need to provide details, like your Azure subscription, resource group, region, instance name, and pricing tier. See Figure 3-4.

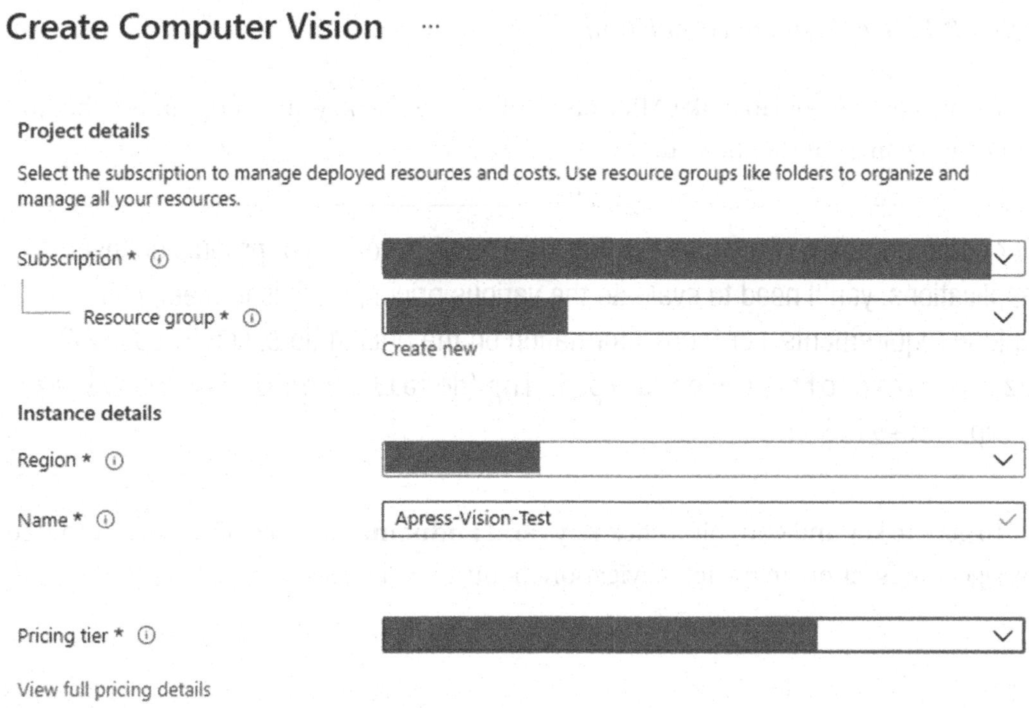

Create Computer Vision ⋯

Project details

Select the subscription to manage deployed resources and costs. Use resource groups like folders to organize and manage all your resources.

Subscription * ⓘ

Resource group * ⓘ

Create new

Instance details

Region * ⓘ

Name * ⓘ   Apress-Vision-Test

Pricing tier * ⓘ

View full pricing details

Figure 3-4. *Setting up Computer Vision*

Follow the on-screen instructions, and complete the process to create the Computer Vision instance. It will take some time. Once the deployment is completed, click **Go to resource**, as shown in Figure 3-5.

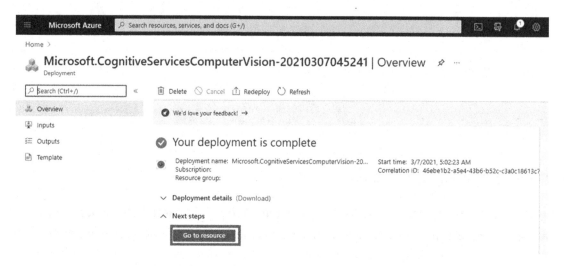

Figure 3-5. *Deployment completion*

You will not be able to make API calls until you get the key and endpoint, so that the application can be authenticated.

For demo purposes, we chose the Free Tier pricing model. For production level applications, you'll need to evaluate the various pricing models to meet your unique requirements. For more information on the pricing tiers, see `https://azure.microsoft.com/en-us/pricing/details/cognitive-services/computer-vision/`.

To get the key and endpoint, click **Keys and Endpoint** available under the Resource Management section, in the left navigation menu. See Figure 3-6.

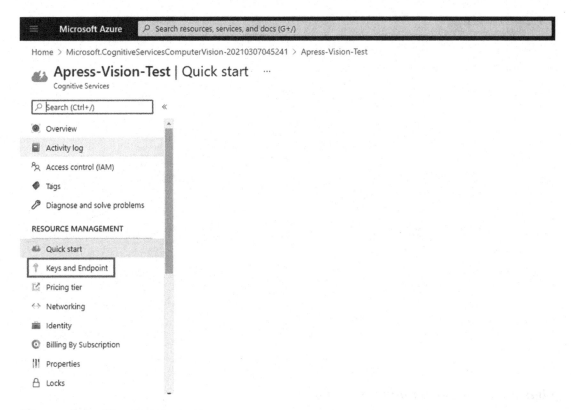

Figure 3-6. *The Quick start page*

From the Keys and Endpoint page, write down your Key 1, Key 2, and Endpoint IDs, to use the Computer Vision resource. See Figure 3-7. From this screen, you can also regenerate the keys.

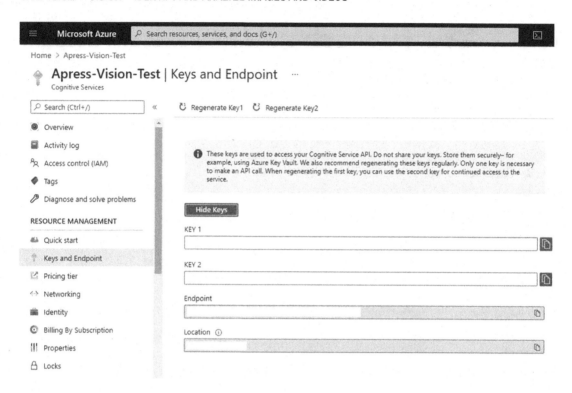

Figure 3-7. *The Keys and Endpoint page*

Working with APIs

Until now, we have created a Computer Vision resource and have set up the endpoint with keys. Now, it's time to work with the APIs. Azure portal provides us multiple ways to test the API, as follows:

1. **API console** – This method to test the API requires a minimal effort. Using the API console, a developer can pass the values to the API, by using the endpoint, and retrieve the results. From the API console, choose the API by clicking the specific region. See Figure 3-8.

Select the testing console in the region where you created your resource:

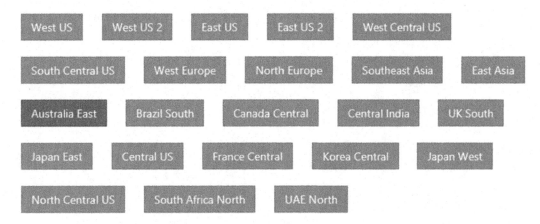

Figure 3-8. *Selecting the API console region*

Select the **Analyze Image** resource, and then provide the required information, as shown in Figure 3-9. Then click **Send**. We tested a random image with this URL: `https://azurecomcdn.azureedge.`
`net/cvt-caf9b3609b1d754524c718b4cde399fda4ea781184fcff2c`
`2e29fbbded7c0ae5/images/shared/cognitive-services-demos/`
`analyze-image/analyze-2-thumbnail.jpg`.

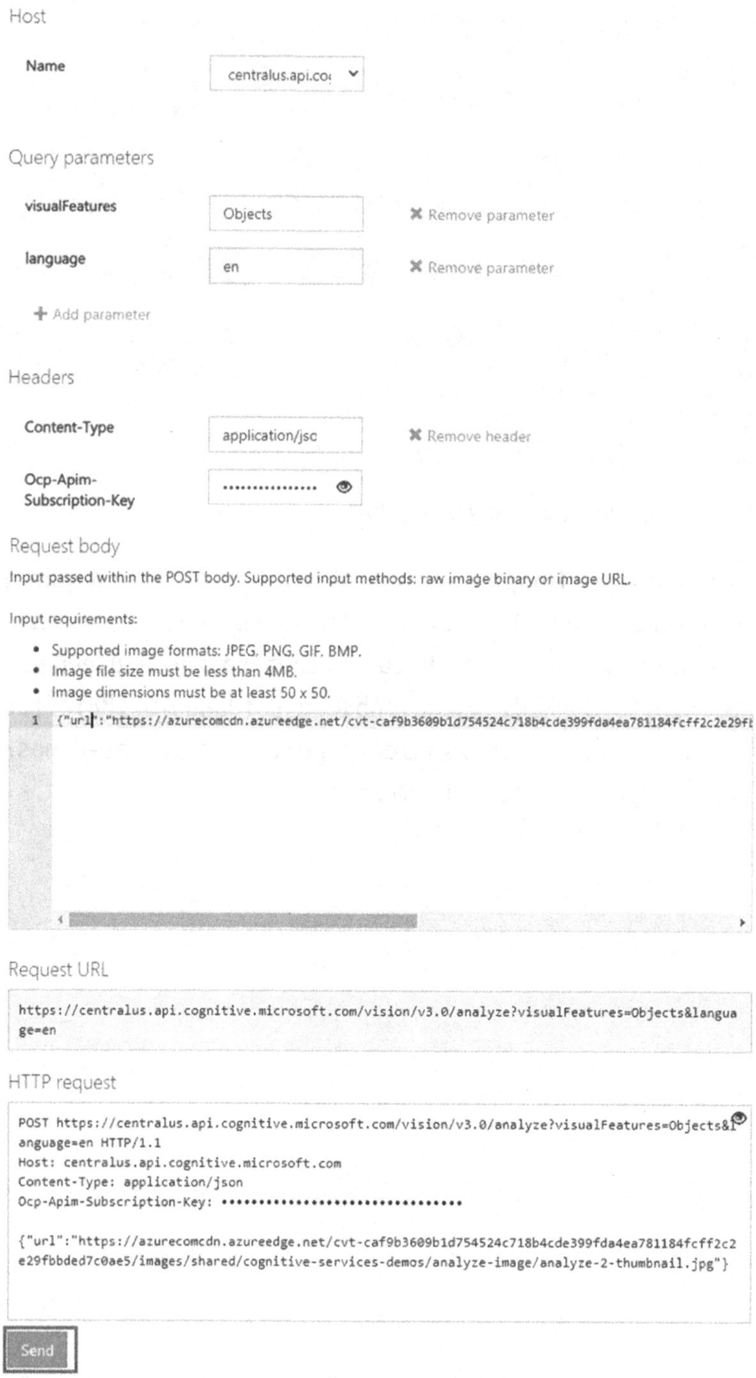

Host

Name

centralus.api.co: ∨

Query parameters

visualFeatures

Objects

✖ Remove parameter

language

en

✖ Remove parameter

＋ Add parameter

Headers

Content-Type

application/jsc

✖ Remove header

Ocp-Apim-
Subscription-Key

················ 👁

Request body

Input passed within the POST body. Supported input methods: raw image binary or image URL.

Input requirements:

- Supported image formats: JPEG, PNG, GIF, BMP.
- Image file size must be less than 4MB.
- Image dimensions must be at least 50 x 50.

```
1  {"url":"https://azurecomcdn.azureedge.net/cvt-caf9b3609b1d754524c718b4cde399fda4ea781184fcff2c2e29f
```

Request URL

```
https://centralus.api.cognitive.microsoft.com/vision/v3.0/analyze?visualFeatures=Objects&langua
ge=en
```

HTTP request

```
POST https://centralus.api.cognitive.microsoft.com/vision/v3.0/analyze?visualFeatures=Objects&l
anguage=en HTTP/1.1
Host: centralus.api.cognitive.microsoft.com
Content-Type: application/json
Ocp-Apim-Subscription-Key: ································

{"url":"https://azurecomcdn.azureedge.net/cvt-caf9b3609b1d754524c718b4cde399fda4ea781184fcff2c2
e29fbbded7c0ae5/images/shared/cognitive-services-demos/analyze-image/analyze-2-thumbnail.jpg"}
```

Send

Figure 3-9. *Sending a request to analyze an image*

The previous request will return the response shown in Listing 3-1.

Listing 3-1. Output from the image analysis API

```
csp-billing-usage: CognitiveServices.ComputerVision.
Objects=1,CognitiveServices.ComputerVision.Transaction=1
x-envoy-upstream-service-time: 844
apim-request-id: 6ddc4b00-4a3f-438b-876c-54800ae6250d
Strict-Transport-Security: max-age=31536000; includeSubDomains; preload
x-content-type-options: nosniff
Date: Sun, 07 Mar 2021 00:36:13 GMT
Content-Length: 281
Content-Type: application/json; charset=utf-8

{
  "objects": [{
    "rectangle": {
      "x": 0,
      "y": 47,
      "w": 34,
      "h": 125
    },
    "object": "person",
    "confidence": 0.615
  }, {
    "rectangle": {
      "x": 69,
      "y": 35,
      "w": 72,
      "h": 138
    },
    "object": "person",
    "confidence": 0.807
  }],
```

```
"requestId": "6ddc4b00-4a3f-438b-876c-54800ae6250d",
"metadata": {
  "height": 175,
  "width": 175,
  "format": "Jpeg"
}
}
```

2. **Testing from the Computer Vision feature page** – To analyze a
   test image (for demo purposes), you can use the feature page to
   get the visual results: https://azure.microsoft.com/en-in/
   services/cognitive-services/computer-vision/#features/.
   For demo purposes, let's use the same image that we tested in the
   API console window. We will get the results shown in Figure 3-10.

Figure 3-10. *Image analysis*

The analysis data is shown in Listing 3-2.

Listing 3-2. Identification of the object

```
[
  {
    "rectangle": {
      "x": 6,
      "y": 390,
      "w": 48,
      "h": 40
    },
    "object": "footwear",
    "confidence": 0.513
  },
  {
    "rectangle": {
      "x": 104,
      "y": 104,
      "w": 127,
      "h": 323
    },
    "object": "person",
    "confidence": 0.763
  },
  {
    "rectangle": {
      "x": 174,
      "y": 236,
      "w": 113,
      "h": 74
    },
```

```
    "object": "Laptop",
    "parent": {
      "object": "computer",
      "confidence": 0.56
    },
    "confidence": 0.553
  },
  {
    "rectangle": {
      "x": 351,
      "y": 331,
      "w": 154,
      "h": 99
    },
    "object": "seating",
    "confidence": 0.525
  },
  {
    "rectangle": {
      "x": 0,
      "y": 101,
      "w": 174,
      "h": 329
    },
    "object": "person",
    "confidence": 0.855
  },
  {
    "rectangle": {
      "x": 223,
      "y": 99,
      "w": 199,
      "h": 322
    },
```

```
      "object": "person",
      "confidence": 0.725
  },
  {
      "rectangle": {
         "x": 154,
         "y": 191,
         "w": 387,
         "h": 218
      },
      "object": "seating",
      "confidence": 0.679
  },
  {
      "rectangle": {
         "x": 111,
         "y": 275,
         "w": 264,
         "h": 151
      },
      "object": "table",
      "confidence": 0.601
  }
]
```

3. **Integration/making a call from within the application** – This method is well used. You call the APIs using the API key, which is generated from the Keys and Endpoint page. For our demo purposes, we have a small amount code that's written in C#.

We used Visual Studio 2019 Community Edition. The steps mentioned will remain the same for all the code examples we discussed in this chapter.

Open Visual Studio, and then click **Create a new project**. On the Create a new project page, click **Console App (.NET Core)**, as shown in Figure 3-11. Then, click **Next**.

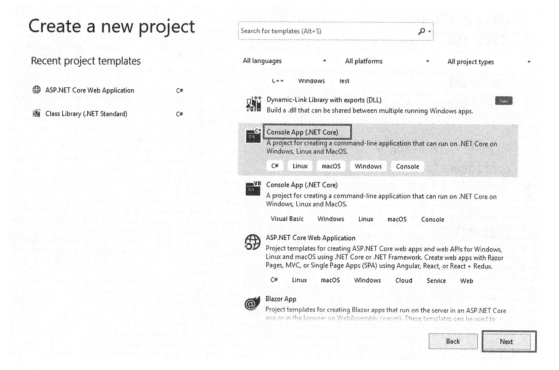

Figure 3-11. *Create a new project*

Name the project, enter the project location path, and provide a valid solution name. Then, click **Create**, as shown in Figure 3-12.

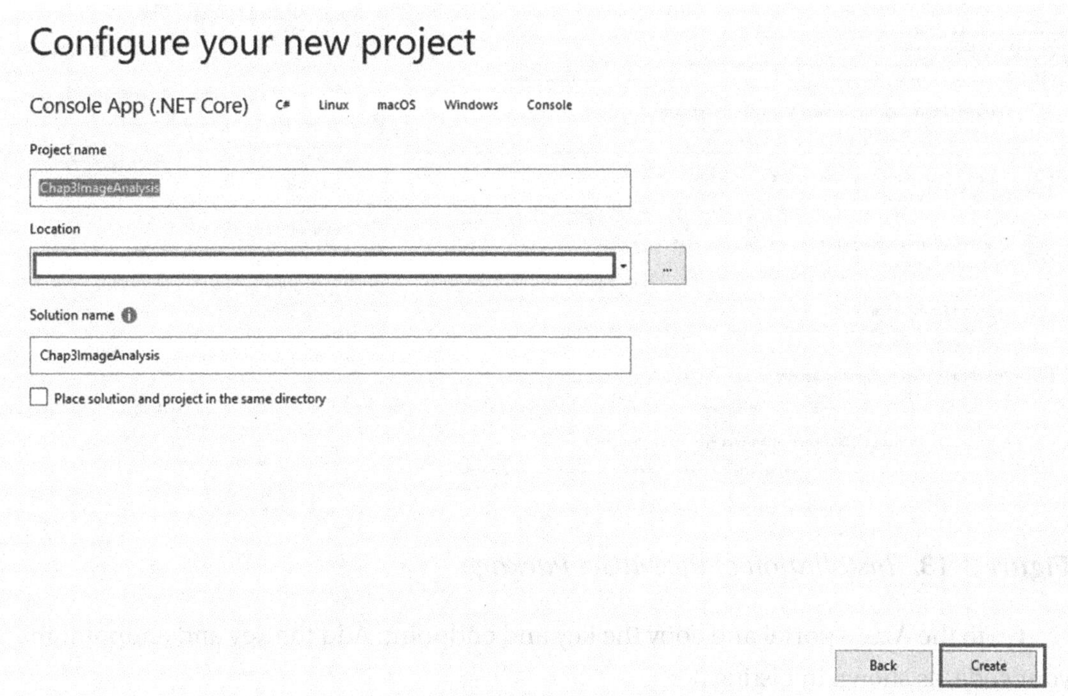

Configure your new project

Console App (.NET Core)   C#   Linux   macOS   Windows   Console

Project name

Chap3ImageAnalysis

Location

Solution name ⓘ

Chap3ImageAnalysis

☐ Place solution and project in the same directory

Back   Create

Figure 3-12. *Configure your new project*

Right-click the project name from Solution Explorer, and then
click **Manage NuGet Packages**. From the NuGet Package
Manager, search for **Microsoft.Azure.CognitiveServices.Vision.
ComputerVision** in the Browse tab. Make sure you check the
Include prerelease check box. In the right pane, select the latest
version, and then click **Install** to the right of it. See Figure 3-13.

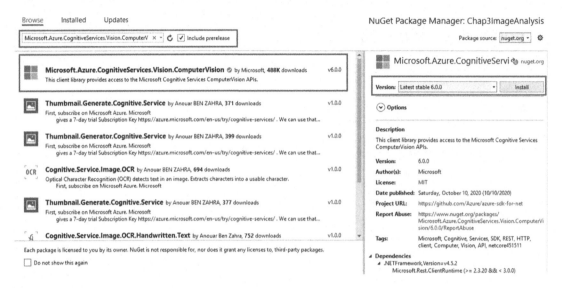

Figure 3-13. *Installation of the NuGet Package*

Go to the Azure portal and copy the key and endpoint. Add the key and endpoint in your code, as shown in Listing 3-3.

Listing 3-3. Adding the Computer Vision subscription key and endpoint

```
static string subscriptionKey = "SUBSCRIPTION_KEY_GOES_HERE";
static string endpoint = "ENDPOINT_GOES_HERE";
```

The subscription is required to validate the requests to the endpoint. You can revisit the previous section and get the subscription key and endpoint.

Set up the AuthenticatedClient. Write the statement shown in Listing 3-4.

Listing 3-4. Creating the client

```
ComputerVisionClient client = AuthenticatedClient(endpoint,
subscriptionKey);
```

We passed an endpoint and a valid subscription key to create an authenticated client. You may see that Visual Studio is throwing an error about the missing namespace. Add the proposed namespace, as shown in Figure 3-14.

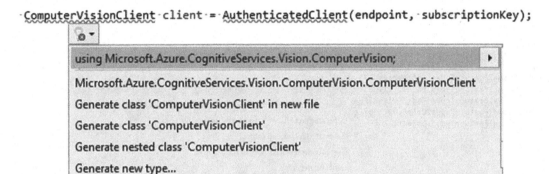

Figure 3-14. *Adding the missing namespace*

The AuthenticatedClient method creates a valid and authenticated client, by validating ApiKeyServiceClientCredentials. See Listing 3-5.

Listing 3-5. Authenticating a client

```
private static ComputerVisionClient AuthenticatedClient(string endpoint,
string subscriptionKey)
{
    ApiKeyServiceClientCredentials clientCredentials = new ApiKeyService
    ClientCredentials(subscriptionKey);
    return new ComputerVisionClient(clientCredentials) { Endpoint = endpoint };
}
```

To add all the features that you are required to analyze, simply create a list of the VisualFeatureTypes model. See Listing 3-6.

Listing 3-6. VisualFeatures to be analyzed

```
List<VisualFeatureTypes?> featuresToBeAnalyzed = new
List<VisualFeatureTypes?>()
{
    VisualFeatureTypes.Categories, VisualFeatureTypes.Description,
    VisualFeatureTypes.Faces, VisualFeatureTypes.ImageType,
    VisualFeatureTypes.Tags, VisualFeatureTypes.Adult,
    VisualFeatureTypes.Color, VisualFeatureTypes.Brands,
    VisualFeatureTypes.Objects
};
```

Add the missing namespace, as shown in Figure 3-15.

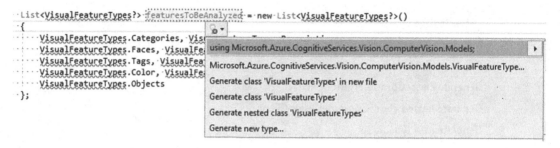

Figure 3-15. *The missing namespace*

Make a request to the Vision API and get the result. The response is the
ImageAnalysis model object, as shown in Listing 3-7.

Listing 3-7. Making a request

```
ImageAnalysis analysisData = await client.AnalyzeImageAsync(imgURL,
featuresToBeAnalyzed);
```

In this short demo, we analyzed every possible feature to check the image
description, type, objects, and so on. The final result of our demo image (`https://docs.`
`microsoft.com/en-us/learn/wwl-data-ai/analyze-images-computer-vision/media/`
`woman-roof.png`) should look like Figure 3-16.

```
Analyzing the image woman-roof.png...

Analysis at a glance:
a woman leaning on a wall with confidence 0.43398770689964294

Image Categories:
people_ with confidence 0.8203125

Image Tags with confidence score:
person 0.9911453723907471
human face 0.9768654108047485
smile 0.9411433935165405
clothing 0.9373063445091248
girl 0.9336122870445251
woman 0.8491367101669312

Identified Objects:
person with confidence 0.883 at location 131, 298, 18, 197

Faces:
A Female of age 27 at location 195, 195, 87, 87

Image Adult contents, if any:
Has adult content: False with confidence 0.023010555654764175
Has racy content: False with confidence 0.05381641536951065
```

Figure 3-16. *The image analysis result*

Identifying a Face

The Computer Vision API is also helpful to identify the human faces in an image. This will analyze the image and return different kinds of data, based on the identified faces.

There are around 27 predefined landmark points that are used to detect human faces. Computer Vision uses these predefined landmarks while processing the images. Refer to the image, `https://docs.microsoft.com/en-us/azure/cognitive-services/face/images/landmarks.1.jpg`, for a pictorial view of these landmarks.

The Face API provides facial analysis. You can perform this analysis to do the following:

- **Detect a face in an image** – It provides the facility to detect a human face with analysis data, by extracting face-related attributes, like a head pose, the person's gender, age, emotion, facial hair, and glasses.

- **Find similar faces in an image** – You might want to search from your database, to find a human face. This API helps you identify a face in an image. This API also further narrows down the search and provides two ways to identify the faces:

  - **matchPerson mode** – It first matches and identifies the same person, and then it returns the match.

  - **matchFace mode** – It returns the matches and ignores the same person's face. This API does not consider whether a face belongs to a particular person, but it finds all the face matches. In other words, it returns the similar matches of faces that may or may not belong to the same person.

- **Identify faces** – Imagine the feature of auto-tagging human faces, where the system automatically tags the image with a human identity (a name or any symbolic identifier) that is then stored in a database. This API helps you identify specific faces from a stored database of images.

To get started with facial recognition, we need to set up the Face API. To do so, follow the same steps that we followed in the previous section, to set up the Vision API. Search for **Face**, and then set up the service by following the instructions.

Note   If you have a very limited requirement to identify the face from an image, then you can do it with the help of the Vision API, using the optional `visualFeatures parameter`[1]. In case you must deal with complex scenarios (where you want to collect all the data related to the facial analysis), then we recommend you use the Face API.

---

[1]https://westus.dev.cognitive.microsoft.com/docs/services/56f91f2d778daf23d8ec6739/operations/56f91f2e778daf14a499e1fa

Make sure you have set up your key and endpoint as well. The Face resource screen should look like Figure 3-17, after a successful setup.

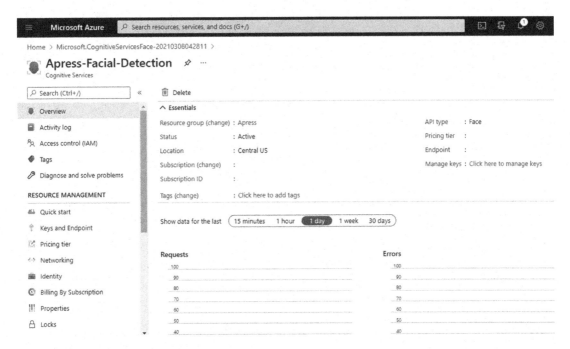

Figure 3-17. *The resource overview screen*

Test Using the API Console

We have set up our Face API. Let's now test it, by using the API console. You need to open the API console from the Azure portal, by selecting the region where you have created your resource. For demo purposes, we created a resource in the Central US region. We will test the API from this region only. We are using a very simple API called Face Detect, with a random image (`https://docs.microsoft.com/en-us/learn/wwl-data-ai/analyze-images-computer-vision/media/woman-roof.png`). We are also analyzing the image for the following facial attributes: age and gender. See Figure 3-18.

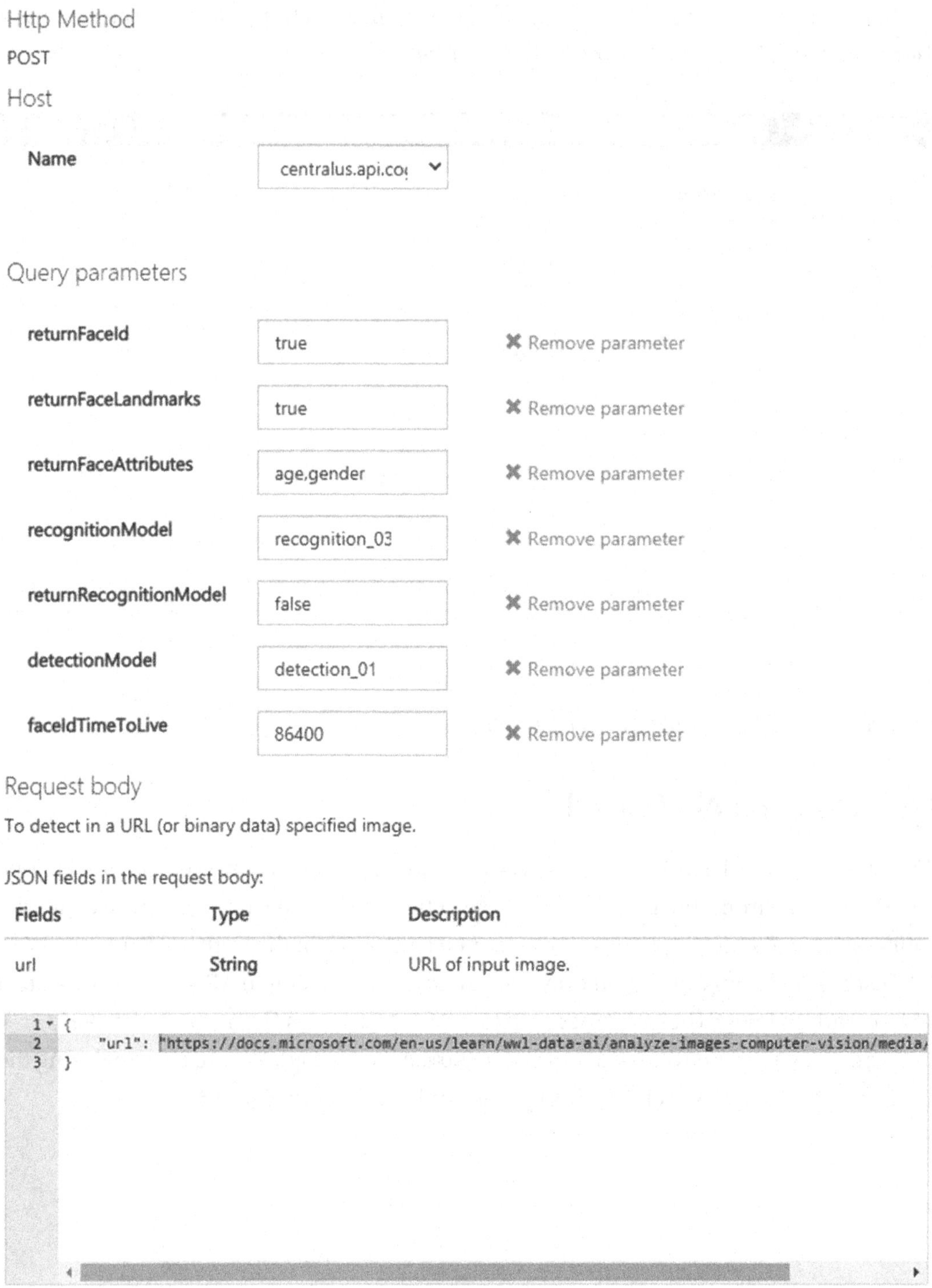

Http Method

POST

Host

Name		
	centralus.api.co	˅

Query parameters

returnFaceId	true	✖ Remove parameter
returnFaceLandmarks	true	✖ Remove parameter
returnFaceAttributes	age,gender	✖ Remove parameter
recognitionModel	recognition_03	✖ Remove parameter
returnRecognitionModel	false	✖ Remove parameter
detectionModel	detection_01	✖ Remove parameter
faceIdTimeToLive	86400	✖ Remove parameter

Request body

To detect in a URL (or binary data) specified image.

JSON fields in the request body:

Fields	Type	Description
url	String	URL of input image.

```
1 ▾ {
2        "url": "https://docs.microsoft.com/en-us/learn/wwl-data-ai/analyze-images-computer-vision/media/
3    }
```

Figure 3-18. *Requesting the Face Detect API*

Note There are prebuilt detection models to identify or analyze the human faces for the Face API – detection_01, detection_02, and detection_03. Make sure you use the same detection model while comparing or identifying similar human faces. In the demo example, we used detection_01 model.

If your request is valid and executed, it will return similar results as what is shown in Listing 3-8.

Listing 3-8. Detecting a face analysis response

```
[{
  "faceId": "0edfbefe-a73a-4143-b6bb-f722d39f97db",
  "faceRectangle": {
    "top": 45,
    "left": 195,
    "width": 42,
    "height": 42
  },
  "faceLandmarks": {
    "pupilLeft": {
      "x": 204.3,
      "y": 59.6
    },
    "pupilRight": {
      "x": 222.5,
      "y": 54.8
    },
    "noseTip": {
      "x": 220.4,
      "y": 66.9
    },
    "mouthLeft": {
      "x": 209.6,
      "y": 78.4
    },
```

```
    "mouthRight": {
      "x": 226.3,
      "y": 74.1
    },
    "eyebrowLeftOuter": {
      "x": 193.1,
      "y": 57.8
    },
    "eyebrowLeftInner": {
      "x": 210.0,
      "y": 51.9
    },
    "eyeLeftOuter": {
      "x": 200.3,
      "y": 60.8
    },
    "eyeLeftTop": {
      "x": 203.4,
      "y": 58.8
    },
    "eyeLeftBottom": {
      "x": 203.8,
      "y": 60.9
    },
    "eyeLeftInner": {
      "x": 207.3,
      "y": 59.3
    },
    "eyebrowRightInner": {
      "x": 217.4,
      "y": 50.5
    },
    "eyebrowRightOuter": {
      "x": 226.6,
      "y": 46.7
    },
```

```
"eyeRightInner": {
  "x": 219.2,
  "y": 55.7
},
"eyeRightTop": {
  "x": 221.9,
  "y": 53.9
},
"eyeRightBottom": {
  "x": 222.4,
  "y": 55.7
},
"eyeRightOuter": {
  "x": 224.9,
  "y": 53.9
},
"noseRootLeft": {
  "x": 211.4,
  "y": 58.3
},
"noseRootRight": {
  "x": 217.3,
  "y": 56.6
},
"noseLeftAlarTop": {
  "x": 213.5,
  "y": 65.5
},
"noseRightAlarTop": {
  "x": 221.4,
  "y": 63.0
},
"noseLeftAlarOutTip": {
  "x": 211.9,
  "y": 69.3
},
```

```
    "noseRightAlarOutTip": {
      "x": 225.1,
      "y": 65.4
    },
    "upperLipTop": {
      "x": 220.3,
      "y": 73.8
    },
    "upperLipBottom": {
      "x": 220.6,
      "y": 75.4
    },
    "underLipTop": {
      "x": 221.3,
      "y": 79.1
    },
    "underLipBottom": {
      "x": 222.2,
      "y": 81.5
    }
  },
  "faceAttributes": {
    "gender": "female",
    "age": 25.0
  }
}]
```

Here, the Face API identifies the human face with the following attributes: the gender is female, and her age is 25 years.

Test Using a Demo Page

If you want to analyze the product, the product overview page for Face API provides you with a demo page. To showcase the power of face detection, from the demo page, we used an author profile pic (of Gaurav). The Face API identifies the face in the photo, as shown in Figure 3-19.

Figure 3-19. *Human face detection*

The analysis data is shown in detail, in Listing 3-9. It provides almost all the possible attributes.

Listing 3-9. Face detection analysis data

```
[
  {
    "faceId": "fef6dba7-6d8d-4825-aa08-8417aa29b563",
    "faceRectangle": {
      "top": 83,
      "left": 92,
      "width": 130,
      "height": 130
    },
```

```
"faceAttributes": {
  "hair": {
    "bald": 0.1,
    "invisible": false,
    "hairColor": [
      {
        "color": "black",
        "confidence": 0.99
      },
      {
        "color": "brown",
        "confidence": 0.65
      },
      {
        "color": "gray",
        "confidence": 0.52
      },
      {
        "color": "other",
        "confidence": 0.44
      },
      {
        "color": "blond",
        "confidence": 0.11
      },
      {
        "color": "red",
        "confidence": 0.03
      },
      {
        "color": "white",
        "confidence": 0.0
      }
    ]
  },
```

```
"smile": 0.999,
"headPose": {
  "pitch": -0.8,
  "roll": -1.8,
  "yaw": -8.6
},
"gender": "male",
"age": 41.0,
"facialHair": {
  "moustache": 0.1,
  "beard": 0.1,
  "sideburns": 0.1
},
"glasses": "ReadingGlasses",
"makeup": {
  "eyeMakeup": false,
  "lipMakeup": false
},
"emotion": {
  "anger": 0.0,
  "contempt": 0.0,
  "disgust": 0.0,
  "fear": 0.0,
  "happiness": 0.999,
  "neutral": 0.001,
  "sadness": 0.0,
  "surprise": 0.0
},
"occlusion": {
  "foreheadOccluded": false,
  "eyeOccluded": false,
  "mouthOccluded": false
},
```

```
    "accessories": [],
    "blur": {
      "blurLevel": "medium",
      "value": 0.63
    },
    "exposure": {
      "exposureLevel": "overExposure",
      "value": 0.82
    },
    "noise": {
      "noiseLevel": "medium",
      "value": 0.39
    }
  },
  "faceLandmarks": {
    "pupilLeft": {
      "x": 130.0,
      "y": 120.2
    },
    "pupilRight": {
      "x": 184.2,
      "y": 117.2
    },
    "noseTip": {
      "x": 156.5,
      "y": 153.2
    },
    "mouthLeft": {
      "x": 128.0,
      "y": 180.4
    },
    "mouthRight": {
      "x": 183.9,
      "y": 177.6
    },
```

```
"eyebrowLeftOuter": {
  "x": 110.7,
  "y": 110.2
},
"eyebrowLeftInner": {
  "x": 146.7,
  "y": 110.1
},
"eyeLeftOuter": {
  "x": 120.8,
  "y": 122.1
},
"eyeLeftTop": {
  "x": 128.5,
  "y": 116.5
},
"eyeLeftBottom": {
  "x": 129.1,
  "y": 123.4
},
"eyeLeftInner": {
  "x": 136.9,
  "y": 120.9
},
"eyebrowRightInner": {
  "x": 167.4,
  "y": 109.1
},
"eyebrowRightOuter": {
  "x": 201.1,
  "y": 107.3
},
"eyeRightInner": {
  "x": 176.2,
  "y": 119.3
},
```

```
  "eyeRightTop": {
    "x": 183.1,
    "y": 113.8
  },
  "eyeRightBottom": {
    "x": 184.4,
    "y": 120.6
  },
  "eyeRightOuter": {
    "x": 192.2,
    "y": 117.7
  },
  "noseRootLeft": {
    "x": 148.6,
    "y": 122.2
  },
  "noseRootRight": {
    "x": 164.1,
    "y": 121.8
  },
  "noseLeftAlarTop": {
    "x": 144.6,
    "y": 143.8
  },
  "noseRightAlarTop": {
    "x": 169.5,
    "y": 142.9
  },
  "noseLeftAlarOutTip": {
    "x": 137.6,
    "y": 155.3
  },
  "noseRightAlarOutTip": {
    "x": 175.8,
    "y": 153.2
  },
```

```
      "upperLipTop": {
        "x": 155.1,
        "y": 170.1
      },
      "upperLipBottom": {
        "x": 155.4,
        "y": 175.1
      },
      "underLipTop": {
        "x": 154.8,
        "y": 185.7
      },
      "underLipBottom": {
        "x": 155.5,
        "y": 192.7
      }
    }
  }
]
```

If we look into the data, then we can easily say that this face belongs to a human that is a male of age 41 years, with black hair, who wears reading glasses and has a smile on his face.

Implement It Using Code

With the help of these APIs, it's very easy to integrate APIs in an existing code, or you can write a new application to analyze the faces. In this section, we included a small demo to help you understand the code and the power of the APIs. You can find the complete code in the repository at [URL of the repository].

We created a Visual Studio project. (Please refer to the previous section to recall how we built a Visual Studio project.) Once you've created the project, open NuGet Package Manager and search for the package **Microsoft.Azure.CognitiveServices.Vision.Face**. Click **Install** in the right pane, as shown in Figure 3-20.

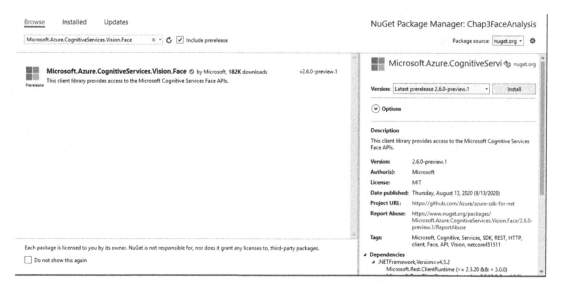

Figure 3-20. *Adding a NuGet package*

Before, you created a client to call the Face API, and you added a name space. See Listing 3-10. This namespace will help you resolve all the references, so you can easily create a face client.

Listing 3-10. Namespace

```
using Microsoft.Azure.CognitiveServices.Vision.Face;
```

Let's create a face client. You need a valid subscription key for this. Write the method to create a client, as shown in Listing 3-11.

Listing 3-11. Creating a face client

```
private static FaceClient AuthenticatedClient(string endpoint, string
subscriptionKey)
{
    ApiKeyServiceClientCredentials clientCredentials = new ApiKeyService
    ClientCredentials(subscriptionKey);
    return new FaceClient(clientCredentials) { Endpoint = endpoint };
}
```

Here, we used subscriptionkey as the client's credentials. Then we provide the endpoint to create the FaceClient.

Let's perform a simple face detection. This will provide an overview of the identified face, by getting its location in the image. Before we write the code to start an analysis, we must add a few namespaces, as shown in Listing 3-12.

Listing 3-12. Face model namespaces

```
using Microsoft.Azure.CognitiveServices.Vision.Face;
using Microsoft.Azure.CognitiveServices.Vision.Face.Models;
```

The model's namespace will let you access the DetectedFace model, which contains the analysis data. We added a simple call to the API by using our face client. We used an image with the URL shown in Listing 3-13. Complete example is available on GitHub (`https://github.com/Apress/hands-on-azure-cognitive-services/tree/main/Chapter%2003/Chap3FaceAnalysis`).

Listing 3-13. Requesting a Face API

```
IList<DetectedFace> faces = await client.Face.DetectWithUrlAsync(url:
imgURL, returnFaceId: true, detectionModel: DetectionModel.Detection02);
```

In this code, Detection02 is a predefined detection model. You can use either Detection01 or Detection02 for our demo example. There are two request methods, DetectWithUrlAsync and DetectWithStreamAsync. The first method detects a face in the image, from the provided URL. The second method (DetectWithStreamAsync) expects a stream of the image. (This is important when you are building a user interface to upload the images.) Listing 3-14 shows how you can retrieve the data from the DetectedFace model list, which is returned from the API.

Listing 3-14. Information about the detected face

```
foreach (var face in faces)
{
    FaceRectangle rect = face.FaceRectangle;
    Console.WriteLine($"Face:{face.FaceId} is located in the image in
    marked point having dimensions: height-{rect.Height} width-{rect.Width}
    which is available from Top-{rect.Top} Left-{rect.Left}.");

}
```

Run the project. The result should be similar to what's shown in Figure 3-21.

```
Chapter-3 Computer facial analysis demo

Analyzing the image woman-roof.png...

Analysis at a glance:
Face:000e4a40-7a7b-4a9b-a934-40deda12e949 is located in the image in marked point having dimensions: height-62 width-45
which is available from Top-32 Left-185.

---------------------------------------------------------

Image analysis is complete.

Press enter to exit...
```

Figure 3-21. *The facial analysis output*

Understanding the Working Behavior of Vision APIs for Video Analysis

The Video Indexer service helps us to provide the video analysis. This automatically extracts the metadata from videos (motion graphics), such as words, written text, faces, speakers, and the identity of celebrities. In the following example, we will use Azure Cognitive Services to analyze video frames from a webcam, in near real time, by using the OpenCVSharp package:

1. The first step is to clone the repository from GitHub[2], as shown in Figure 3-22. You will find both applications in the repository's Windows folder.

---

[2]https://github.com/microsoft/Cognitive-Samples-VideoFrameAnalysis

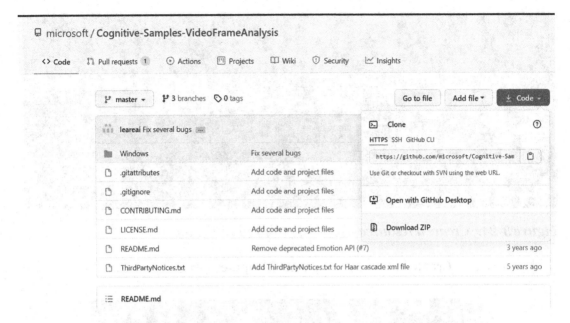

Figure 3-22. *Cognitive samples video frame analysis GitHub repository*

2.   Next, open the Azure portal to create a resource for Computer
     Vision. Click the ***Create a resource*** button, as shown in Figure 3-23.

Figure 3-23. *Creating a resource on the Azure portal*

Search for and select the **Computer Vision** resource type in the search bar.

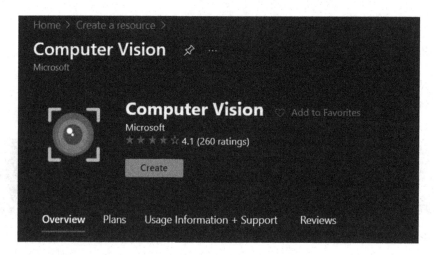

Figure 3-24. *Create a resource – selecting Computer Vision*

Now click the **Create** button to create your Computer Vision resource.

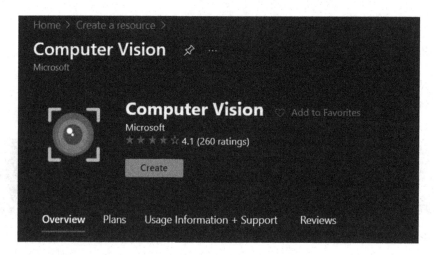

Figure 3-25. *Creating the resource*

This takes you to the Create Computer Vision resource page. Here, you can create the Computer Vision resource by providing the relevant information about your subscription, resource group, and so on (as shown in Figure 3-26).

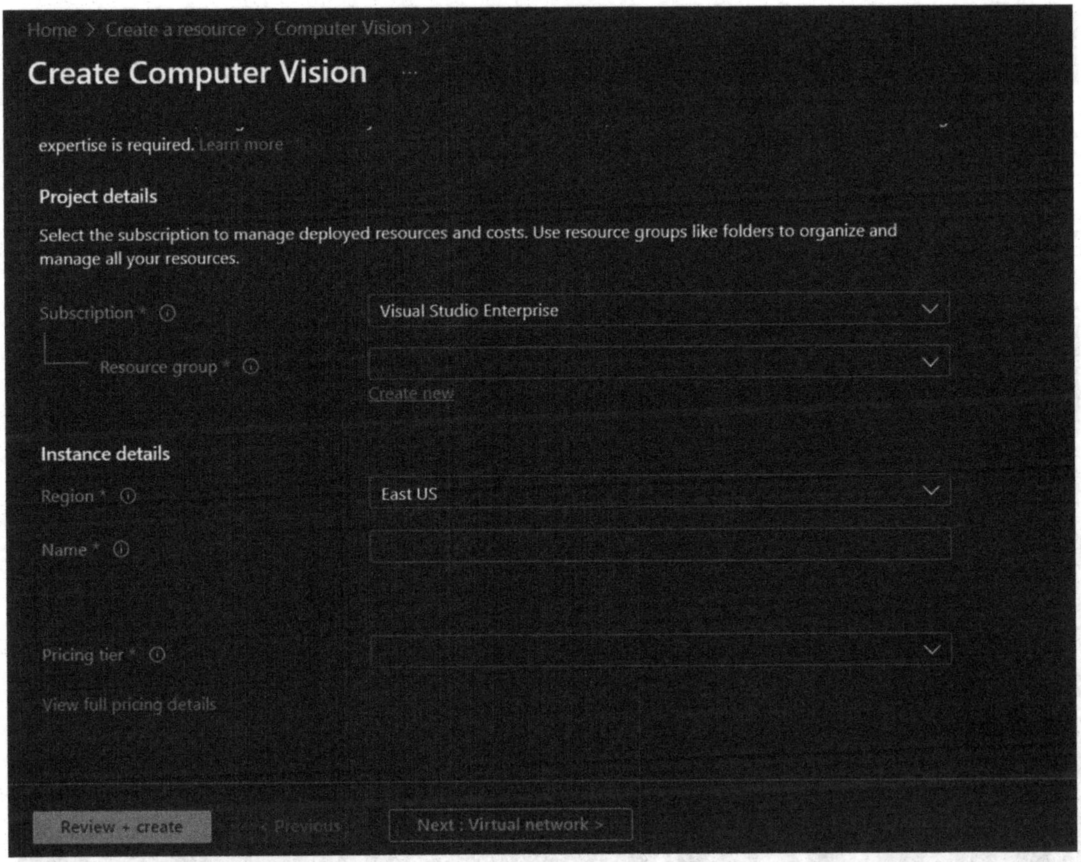

Figure 3-26. *Creating a Computer Vision resource*

Once you have selected and completed the details, click the **Review + create** button (shown in Figure 3-27) to validate the information that you provided.

Figure 3-27. *The Review + create button*

Once you validate the information, click **Create** to start and deploy the Computer Vision resource.

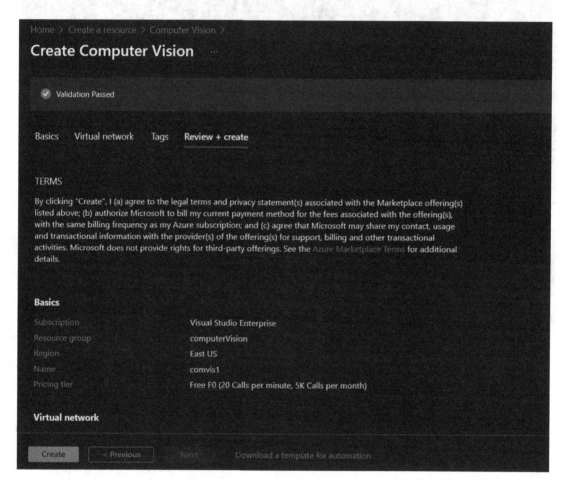

Figure 3-28. *Validating the created resource*

You will then see the notification that the deployment is in progress (as shown in Figure 3-29), after which you will be able to use the service.

Figure 3-29. *The Computer Vision resource deployment in progress*

1. In this step, you will create another resource in the Azure portal. This will be the Face API, which you'll use for face detection. See Figure 3-30.

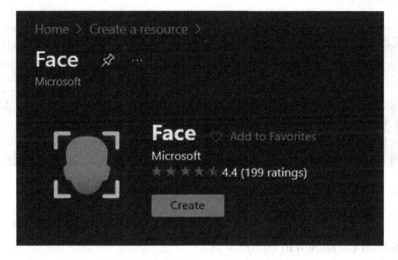

Figure 3-30. *Creating a Face resource*

The rest of the steps are the same for creating a resource, as shown earlier. The Face API deployment will take a couple of minutes. Figure 3-31 shows the Face resource deployment in progress.

Figure 3-31.   *The Face resource deployment in progress*

1. Now that you have created the services, let's open the cloned
    repository from the destination folder in your local machine, by
    using Visual Studio.

    Within Visual Studio, click **File**, **Open**, and **Folder** (as shown in
    Figure 3-32).

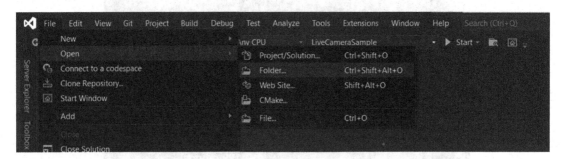

Figure 3-32.   *Opening the solution folder from the downloaded repo*

By opening the solution, you will now see the files in your Solution
Explorer (as shown in Figure 3-33).

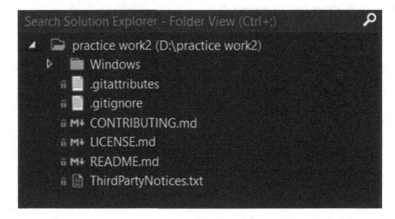

Figure 3-33.   *Opening the solution folder from the downloaded repo*

Expand the Windows folder, and then click the **VideoFrameAnalysis.sln** file (shown in Figure 3-34).

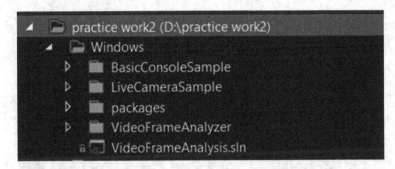

Figure 3-34. *Opening the VideoFrameAnalysis.sln file*

Double-click **LiveCameraSample** to load it.

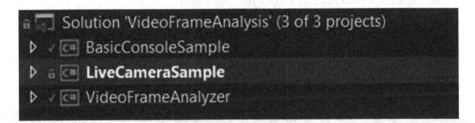

Figure 3-35. *The LiveCameraSample project*

Nuget is the package manager for .NET. To use the camera solution, you will need the Cognitive Services package, which is a library that's used to work with Azure Cognitive Services. Next, right-click the **Solution 'VideoFrameAnalysis'** option, and then click **Manage NuGet Packages for Solution**... (as shown in Figure 3-36).

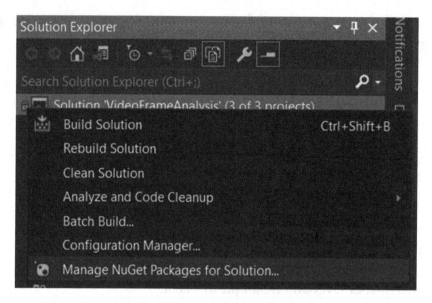

Figure 3-36. *Manage NuGet Packages for Solution*

At this point, try to restore all the required NuGet packages for
the solution, and then validate if they are installed, as shown in
Figure 3-37.

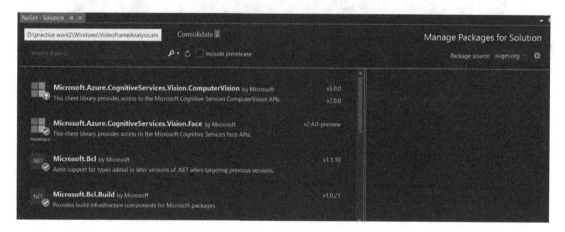

Figure 3-37. *Managing the NuGet Packages for the solution*

As seen in Figure 3-37, the packages do not include the Computer
Vision package. After you restore the packages, Nuget will look like it
does in Figure 3-38. Please ignore the package depreciation warning.

Figure 3-38. *The restored packages*

2. In the Basic Console Sample, the Face API key is hard-coded
    directly in the BasicConsoleSample/Program.cs file. See
    Figure 3-39.

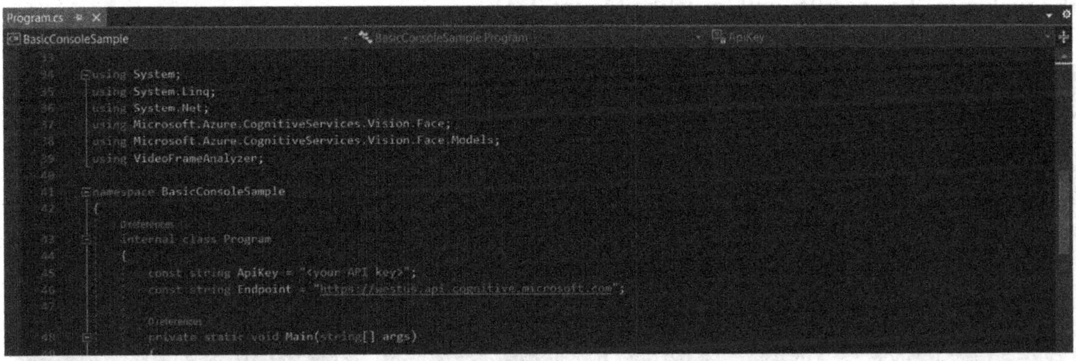

```
using System;
using System.Linq;
using System.Net;
using Microsoft.Azure.CognitiveServices.Vision.Face;
using Microsoft.Azure.CognitiveServices.Vision.Face.Models;
using VideoFrameAnalyzer;

namespace BasicConsoleSample
{
    internal class Program
    {
        const string ApiKey = "<your API key>";
        const string Endpoint = "https://westus.api.cognitive.microsoft.com";

        private static void Main(string[] args)
```

Figure 3-39. *The BasicConsoleSample.Program cs file*

Get the API key and endpoint from the Azure portal, from the Face resources that
you created earlier (as shown in Figure 3-40).

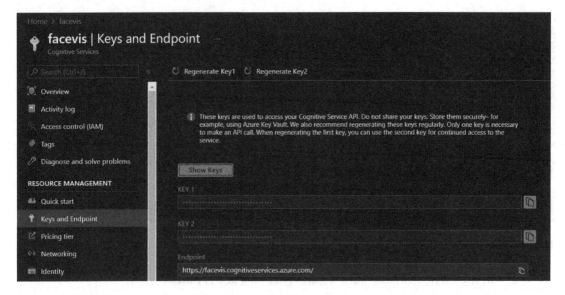

Figure 3-40. *The Face API keys and endpoints on the Azure portal*

Once completed, run the BasicConsoleSample file, which will read the frames from the webcam. The results will look like Figure 3-41.

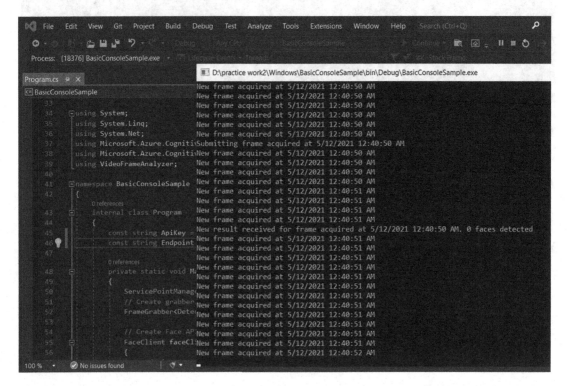

Figure 3-41. *Running the Basic Console Sample and acquiring the frames*

Next, run the LiveCameraSample. Its UI asks you for the API keys and endpoint for both the Face and Vision APIs that we created earlier. See Figure 3-42. Copy and paste the keys, and then click **Save**.

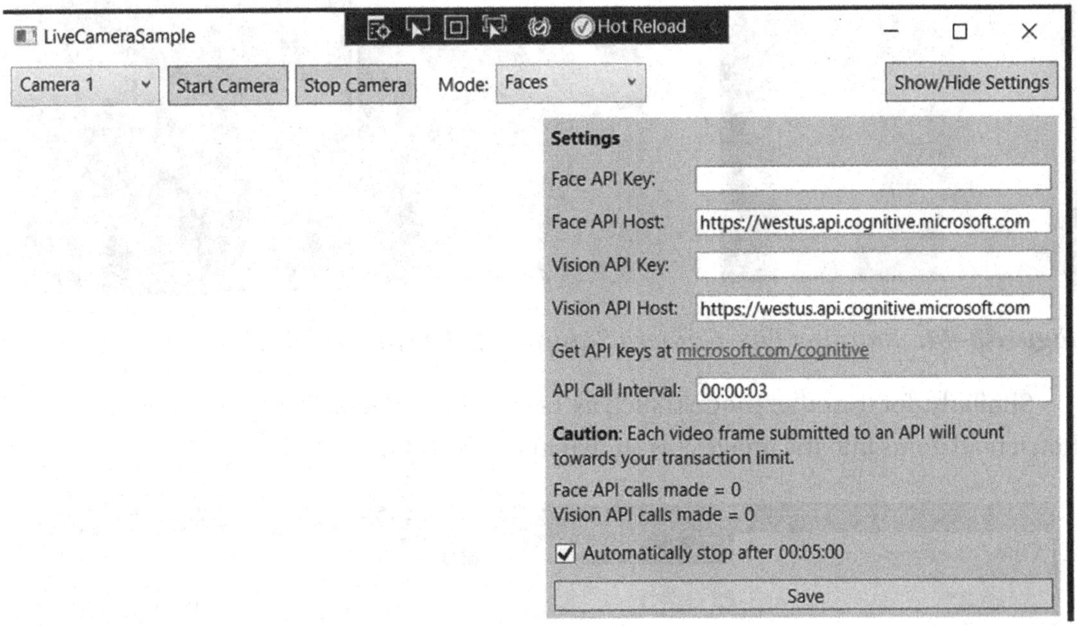

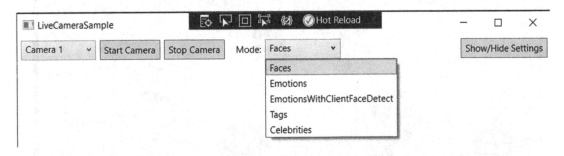

Figure 3-42. *Running Live Camera Sample*

Select a specific mode, and then click **Start Camera**. See Figure 3-43.

Figure 3-43. *Running Live Camera Sample*

Here, we used the Celebrities mode. As expected, the program recognized our picture of Bill Gates (as shown in Figure 3-44).

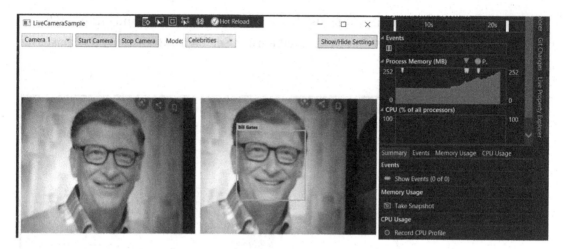

Figure 3-44. *Running Live Camera Sample – Celebrities mode*

Similarly, for the other modes (such as Tags for object detection), you can put household items into the images (as shown in Figure 3-45).

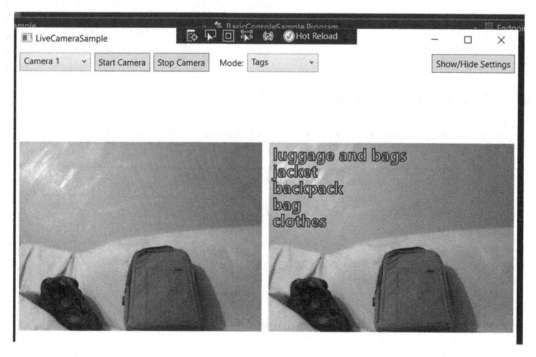

Figure 3-45. *Running Live Camera Sample – Tags mode*

In this example, you can see how you can capture the live camera feeds. You can use Cognitive Services and the Face API to detect faces and objects. In the next example, we will use the Video Indexer service to extract information out of videos.

Microsoft Azure Video Indexer

The Microsoft Azure Video Indexer is an application that's part of Azure Media Services, and it's also built on Cognitive Search and other Azure Cognitive Services (such as Face API, Microsoft Translator, Computer Vision API, and Custom Speech). It provides you with the capability to automatically extract advanced metadata from your video and audio content. To get started, follow these steps:

1. Visit the Video Indexer[3], and then click **Start free** to begin the free trial. See Figure 3-46.

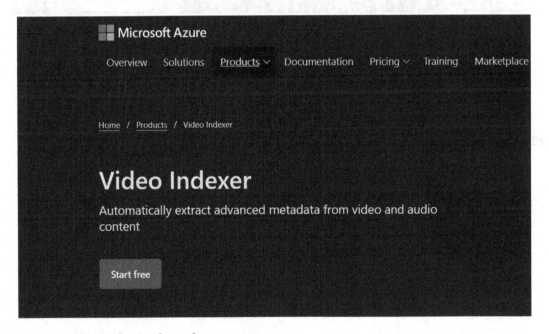

Figure 3-46. *Video Indexer homepage*

---

[3]www.videoindexer.ai/account/login

Log in using your Microsoft account. See Figure 3-47.

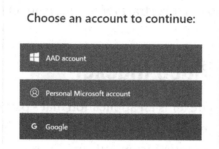

Figure 3-47. *Video Indexer login*

After signing in with your Microsoft account, you will see the screen shown in Figure 3-48. Click the **Media files** film icon on the side bar, to add your video files.

Figure 3-48. *Video Indexer – Extract insights and enhance your content*

Click the **Model customizations** settings icon on the left side bar, to add the person to be identified. In this case, we uploaded the video from one of the talks at ISG Automation Summit London[4] and then selected the person. Click **Add person** from the menu on the right (as shown in Figure 3-49).

---

[4]Dr. Adnan Masood, Chief Architect, UST Global Spoke at ISG Automation Summit, London www.youtube.com/watch?v=2Nv8NZiuotw&ab_channel=UST

Figure 3-49. *Video Indexer – Content model customization*

At this point, you will want to fill in the person's details. The service cannot be used for law enforcement, as Microsoft decided to not sell police its facial recognition technology[5]. Now you will upload the photos of the individual to be identified in the video. See Figure 3-50.

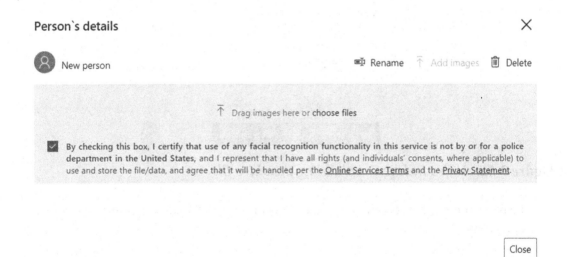

Figure 3-50. *Video Indexer – Person's details*

---

[5]www.washingtonpost.com/technology/2020/06/11/microsoft-facial-recognition/

After the photo is uploaded, person 1 is now added to the model, as shown in Figure 3-51.

Figure 3-51. *Video Indexer – the person is added*

Now go back to the dashboard. Click **Upload** to upload the video that needs to be indexed. Click **browse for a file**, to browse for the file in your system (as shown in Figure 3-52).

Figure 3-52. *Video Indexer – uploading the video file*

The file is then uploaded and analyzed. This process will take some time (as shown in Figure 3-53).

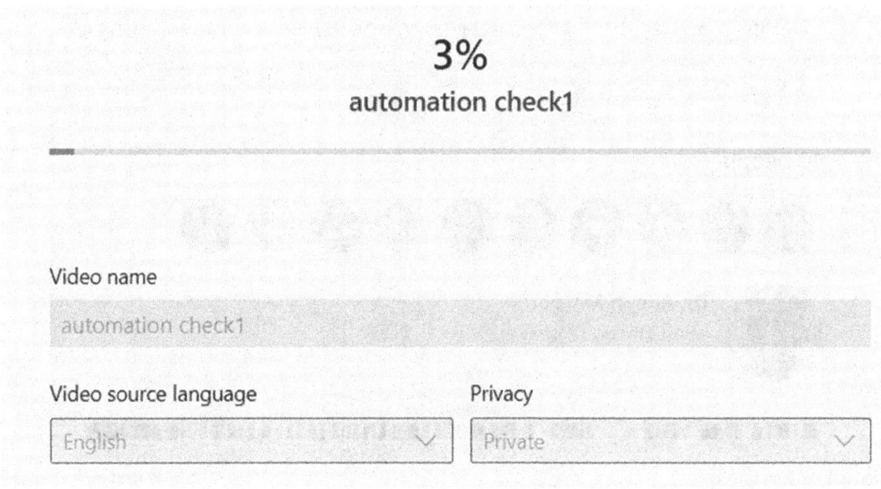

3%
automation check1

Video name

automation check1

Video source language

English

Privacy

Private

Figure 3-53. *Video Indexer – the file upload progress*

Upon successful completion, you will see the analyzed video on the dashboard. Now click the **Play** button on the video (see Figure 3-54).

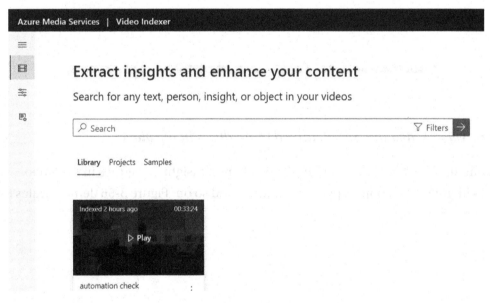

Figure 3-54. *Video Indexer – extracting insights from the video*

As you can see in Figure 3-55, not only did Video Indexer recognize the individual, but it also recognized several other people in the video.

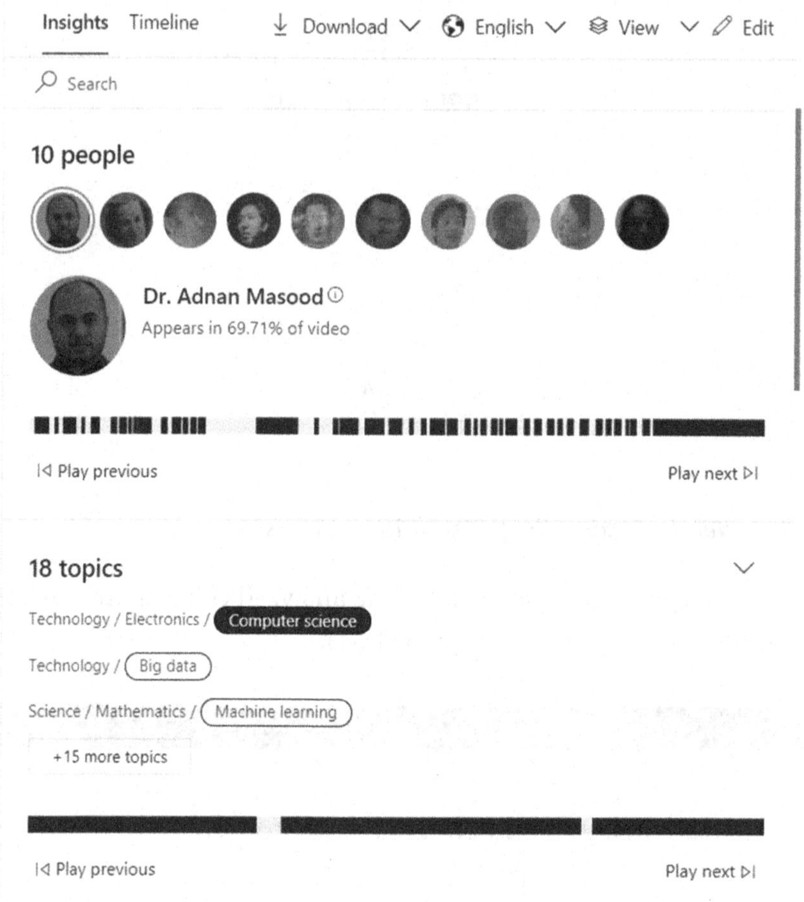

Figure 3-55. *Video Indexer identified individuals in the video*

With this enriched video, we can select different insight categories in the video, such as identifying an object, person, keyword, and so on. Figure 3-56 demonstrates the Timeline report.

Insights **Timeline**   ↓ Download ∨   🌐 English ∨   ⊗ View   ∧ ✎ Edit

🔍 Search

		Show insights	
00:16:55	orator, speech	All	scroll on
00:16:56	standing	Storyboard	
00:16:58	business	Captioning	
00:16:58	KeyFrame #59	Accessibility	
		Monitoring (preview)	

ISG I Automation Summit train ML models mu   Custom view   *uity*

☑	People
☑	Keywords
☑	Labels
☑	Named entities
☑	Emotions
☑	Sentiments
☑	Keyframes
☑	Topics
☑	Scenes

AI infrastructure!used to

text, people, wall, event, presentation, group, ind

00:16:59   *train ML models must be*

I'm sorry Bitcoin, but in the other there are other c

available in perpetuity

00:17:00   sitting, computer

AI infrastructure•used to

train ML models must be   available in perpetuity

00:17:01   wall, human face

00:17:03   *available in perpetuity*

Figure 3-56. *Video Indexer Insights – Timeline*

Video Indexer is a comprehensive and sophisticated video enrichment offering, which can be used in a variety of use cases in media, news, entertainment, journalism, and so on. Here, you have seen some fairly basic capabilities of the service, which can unlock video insights to enrich your videos, to enhance discovery and improve engagement. The feature set includes face detection, celebrity identification, custom face identification, thumbnail extraction, text recognition, object identification, visual content moderation (detecting explicit contents), scene segmentation, shot detection, and so on.

We can only imagine the extent to which you, your company, your clients, and organizations around the world can use this technology!

Summary

In this chapter, we explained the Computer Vision API and various related APIs. With the help of these features, we can identify faces, and we can perform an analysis of any set of images. We can capture the data, and we can make the documents searchable. For motion graphics (videos), the Video Indexer (associated with Azure Media Services) provides us with the facility to identify the audio, words, speakers, written content, and so on.

In the next chapter, we will continue with the discussion of Azure Cognitive Services, and we will take a look at natural language processing (NLP).

CHAPTER 4

Language – Understand Unstructured Text and Models

A smart application can understand a user's input. Users can provide inputs via a keyboard, a mic, or by using any other external devices. These systems are so smart that they can interact with the user.

The aim of this chapter is to get started with natural language processing (NLP) and to create a smart application that can interact with its users. There are a lot of ways to train and prepare the system, so that it can interact with a user.

Throughout this chapter, we will cover the following topics:

- Creating and understanding language models

- Training and enrichment with Cognitive Services

- The Muppet models – transformers for NLP

- Named entity recognition with fine-tuned BERT

- Summary of the language API

Creating and Understanding Language Models

It should come as no surprise that most enterprise data is in an unstructured form. The algorithms that can understand the meaning of words in documents are in huge demand. Since the advent of transformer networks, the field of natural language processing has seen significant improvements in tasks, such as comprehension, questioning and answering, summarization, topic modeling, machine translation,

© Ed Price, Adnan Masood, and Gaurav Aroraa 2021
E. Price et al., *Hands-on Azure Cognitive Services*, https://doi.org/10.1007/978-1-4842-7249-7_4

sentiment analysis, text generation, information retrieval, and relationship extraction. These natural language processing tasks are only as diverse as the document types that we deal with in an enterprise.

Training an enterprise language model requires an understanding of multiple modalities, as well as the nuances associated with a multitude of document types. Most unstructured data comes from emails, wikis (SharePoint, confluence, and so on), and other official documents, such as client agreements, vendor agreements, NDAs, employee contracts, loan and lease documents, investment and compliance documents, amendments, purchase orders, work orders, and so on. These documents vary in nomenclature, jargon, and terminologies, as well as in structure and form. A language model must also deal with legal documents and documents with governance and audit needs, such as term and termination, confidentiality, limitation of liability, indemnification, governing law, assignments, notices, and clauses, such as non-solicitation and noncompete. Extracting such legal clauses can be a challenging task, if applied in a general-purpose setting. A natural language processing pipeline comprises of a variety of steps, as shown in Figure 4-1.

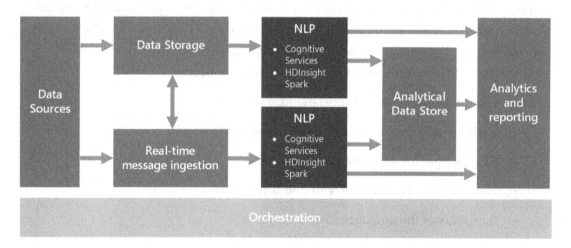

Figure 4-1. *An end-to-end natural language processing pipeline – courtesy of*
`https://docs.microsoft.com/en-us/azure/architecture/data-guide/`
`technology-choices/natural-language-processing`

A typical language pipeline deals with multimodal data formats and storages, as well as with the NLP engine, which is one part of the entire end-to-end life cycle. For instance, one important use case of such a language model may require entity recognition. This could be beyond the typical subject, object, and verb pairs, but instead with business and contractual entities. These include the title of a document, party names and addresses, effective dates, terms and legal clauses, fees, copayments, and so on. Once identified and extracted, you'd analyze these entities, to be presented in the business report and translated into actionable items. For instance, if you would like to know when a particular LLC expires (in order to send an alert to your customer), you can use the extracted filing date and create an alert for 30, 60, and 90 days before the expiration, to let the customer know.

As the industry matures, businesses now look for outcome-driven, end-to-end solutions and complete automated "jobs" (larger groups of tasks), as compared to individual, piecemeal tasks performed by the AI. In the next example, we'll use a demo to show you how this unstructured text enrichment works.

Enrichment in Progress – JFK Files Demo

To demonstrate the power of Cognitive Services and Azure Machine Learning, Microsoft took the complex dataset of the JFK files and put together an impressive demo (see `www.microsoft.com/ai/ai-lab-jfk-files`). In this example, we will show you how the demo works and how you can recreate it for yourself. Figure 4-2 shows the end-to-end enrichment model, starting with the blob storage, where complex files (such as photos, handwriting, and unclassified documents) are stored. Then, Cognitive Search is used to extract information.

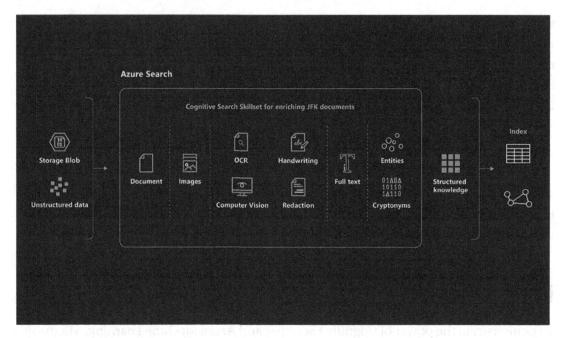

Figure 4-2. *Azure Search for JFK files – courtesy of* `https://github.com/` `microsoft/AzureSearch_JFK_Files`

1. To implement infrastructure as code for Azure solutions, the sample includes Azure Resource Manager templates (commonly known as ARM templates). In this case, you will deploy the ARM template to set up the project on Azure, while you do the rest of the configuration steps. You can find the deployment link at `https://github.com/microsoft/AzureSearch_JFK_Files`, as shown in Figure 4-3.

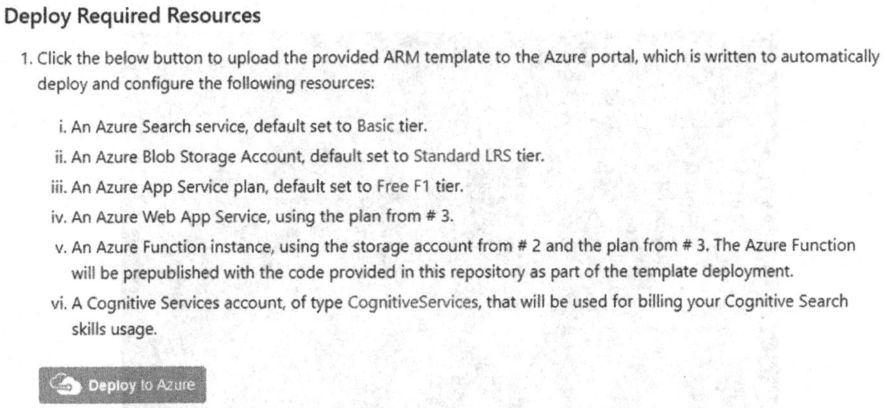

Deploy Required Resources

1. Click the below button to upload the provided ARM template to the Azure portal, which is written to automatically deploy and configure the following resources:

    i. An Azure Search service, default set to Basic tier.

    ii. An Azure Blob Storage Account, default set to Standard LRS tier.

    iii. An Azure App Service plan, default set to Free F1 tier.

    iv. An Azure Web App Service, using the plan from # 3.

    v. An Azure Function instance, using the storage account from # 2 and the plan from # 3. The Azure Function will be prepublished with the code provided in this repository as part of the template deployment.

    vi. A Cognitive Services account, of type CognitiveServices, that will be used for billing your Cognitive Search skills usage.

Figure 4-3. ARM deployment of Azure Search for JFK files – courtesy of https://github.com/microsoft/AzureSearch_JFK_Files

2. One step in deploying the custom template is to create the setup for it. Click the **Deploy to Azure** button. The Custom deployment screen opens, as shown in Figure 4-4. You then select and set the region, resource prefix, hosting plan, search service, and so on, as shown in Figure 4-4.

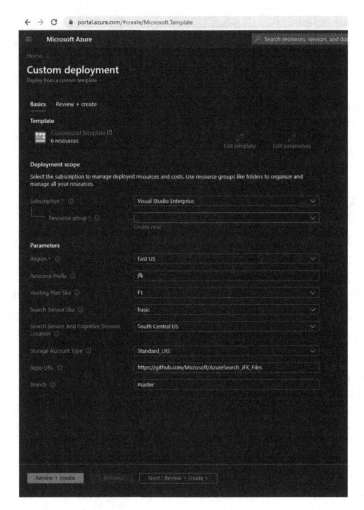

Figure 4-4. *Azure portal Custom deployment*

Click **Create new** and select a resource group, which is a container that holds all the relevant resources. See Figure 4-5.

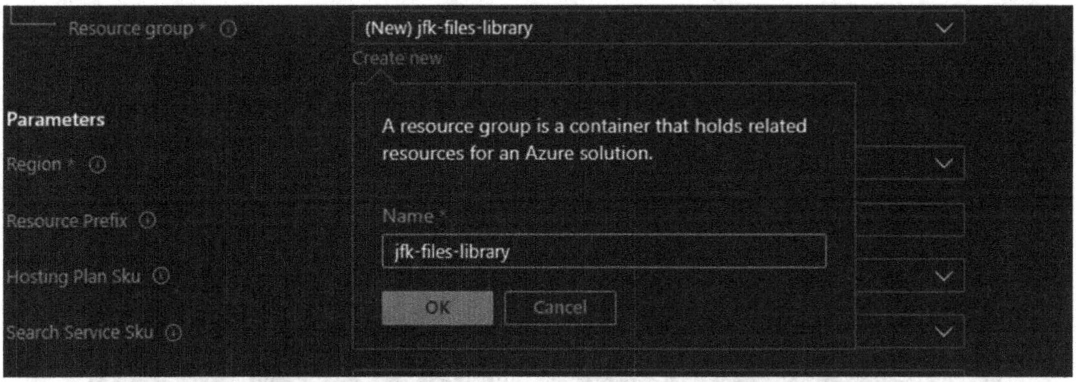

Figure 4-5. *Creating a new resource*

Click **Review + create** to continue, as shown in Figure 4-6.

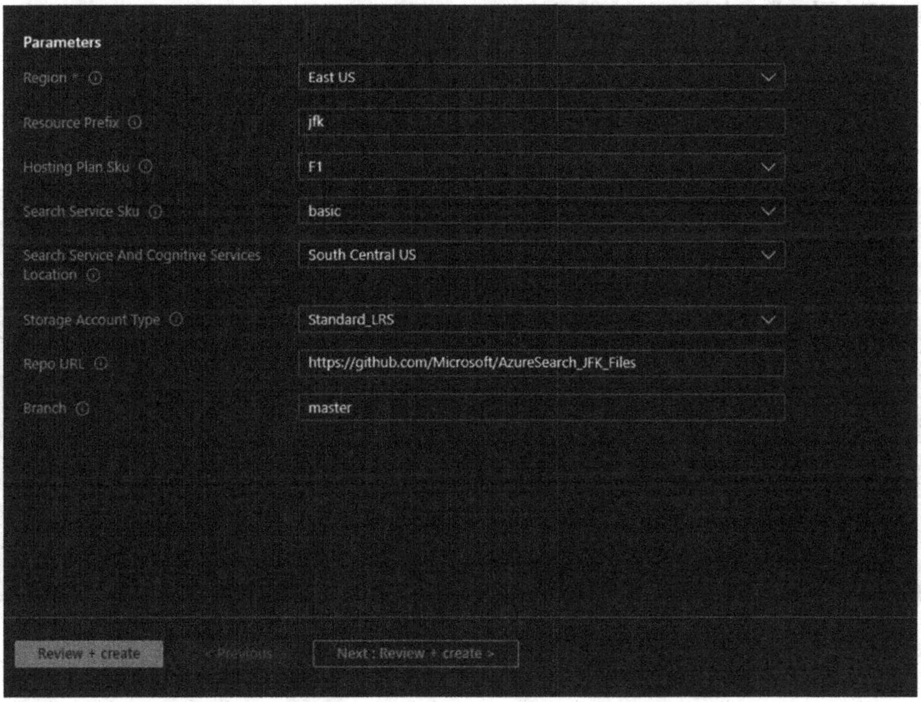

Figure 4-6. *Click Review + create*

Review the subscription information, and then click **Create**, as shown in Figure 4-7.

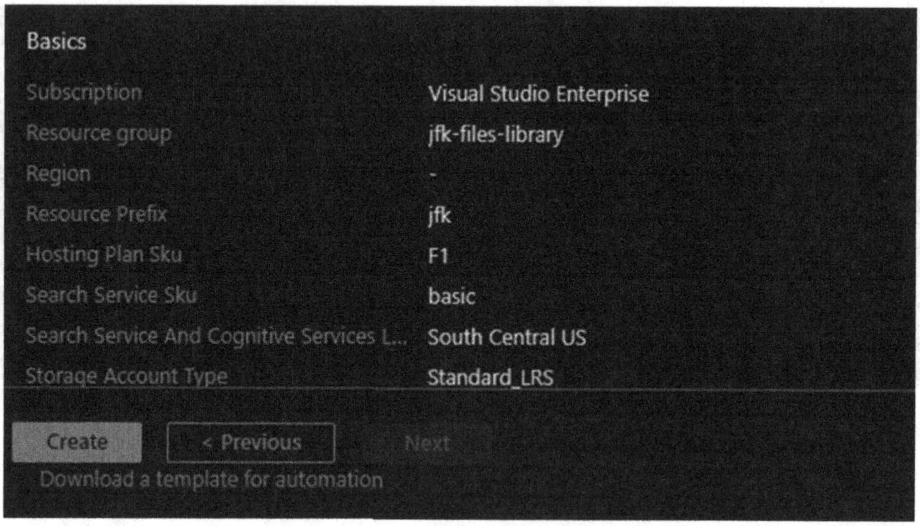

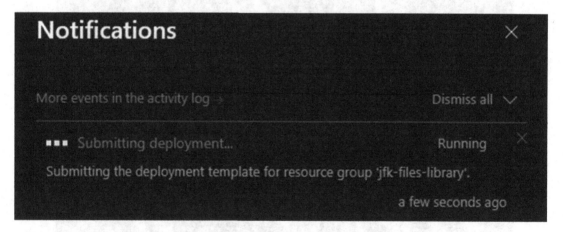

Figure 4-7. *The final step in creating a resource*

Once the deployment is created, you will see the notification that it's submitting a deployment, as shown in Figure 4-8.

Notifications ✕

More events in the activity log →   Dismiss all ⌄

▪▪▪ Submitting deployment...   Running ✕

Submitting the deployment template for resource group 'jfk-files-library'.

a few seconds ago

Figure 4-8. *Notification that the deployment is in progress*

Once the deployment completes, you will see deployment succeeded notification, as shown in the Figure 4-9.

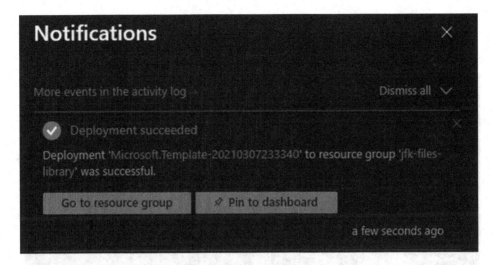

***Figure 4-9.** Deployment succeeded notification*

3. Open Visual Studio (as shown in Figure 4-10), and then clone the
   repository from `https://github.com/microsoft/AzureSearch_`
   `JFK_Files` on GitHub.

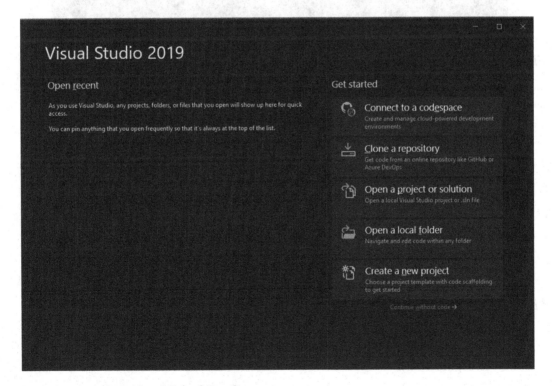

***Figure 4-10.** Opening Visual Studio*

Click the **File** tab, click **Open**, and then click **Project/Solution** (as shown in Figure 4-11).

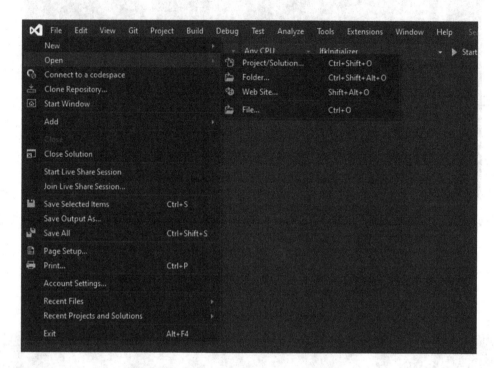

Figure 4-11. *Opening a project*

Once the repository is cloned, open the project directory from the **JfkWebApiSkills** Visual Studio solutions file (as shown in Figure 4-12).

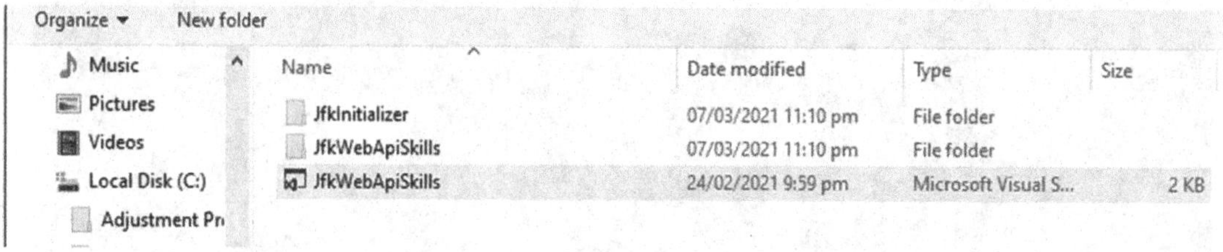

Figure 4-12. *Opening the JfkWebApiSkills solution file*

Next, from the Solution Explorer, edit the **App.config** file (as shown in Figure 4-13).

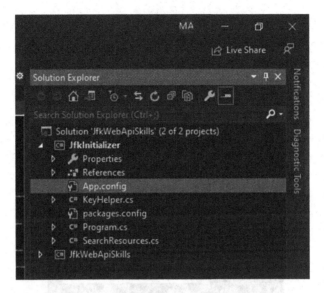

Figure 4-13. *Selecting the App.config file*

Now click the **Outputs** tab in the Azure console, as shown in
Figure 4-14. This tab is available in the deployment screen of the
template.

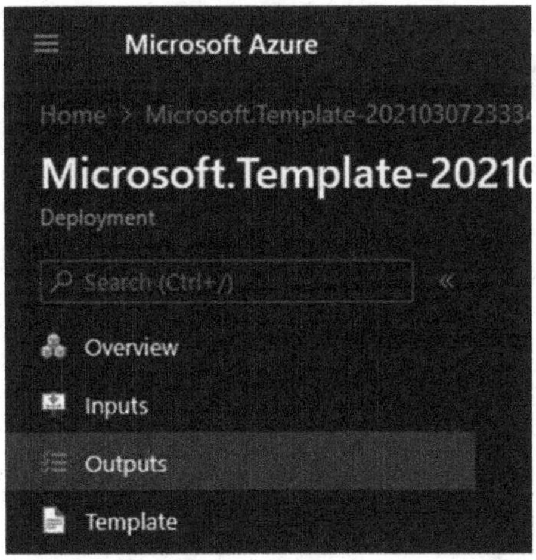

Figure 4-14. *The Outputs tab*

You will then see the template information, with all the keys and strings (as shown in Figure 4-15).

Figure 4-15. *The template Outputs screen*

Copy and paste the values, except the SearchServiceQueryKey, into the config key settings. See Figure 4-16.

```xml
1    <?xml version="1.0" encoding="utf-8"?>
2    <configuration>
3        <startup>
4            <supportedRuntime version="v4.0" sku=".NETFramework,Version=v4.7.2" />
5        </startup>
6        <appSettings>
7            <!--Keys related to your Azure Search instance-->
8            <add key="SearchServiceName" value="jfk-search-service-q66nidjkdjwym" />
9            <add key="SearchServiceApiKey" value="" />
10           <add key="SearchServiceQueryKey" value="" />
11
12           <!--Keys related to your Cognitive Services account-->
13           <add key="CognitiveServicesAccountKey" value="" />
14
15           <!--Keys related to your storage account for the image-store custom skill-->
16           <add key="BlobStorageAccountConnectionString" value="" />
17
18           <!--Keys related to your Azure Web App instance-->
19           <add key="AzureWebAppSiteName" value="" />
20           <add key="AzureWebAppUsername" value="" />
21           <add key="AzureWebAppPassword" value="" />
22
23           <!--Keys related to your Azure Function instance-->
24           <add key="AzureFunctionSiteName" value="" />
25           <add key="AzureFunctionUsername" value="" />
26           <add key="AzureFunctionPassword" value="" />
27
28           <!--Location of sample JFK file documents that we are providing to you.-->
29           <add key="JFKFilesBlobStorageAccountConnectionString" value="SharedAccessSignature=st=2019-05-02T21%3A30%3A00Z&amp
30           <add key="JFKFilesBlobContainerName" value="jfkfiles" />
31
32           <!--Configurable names, feel free to change these if you like-->
33           <add key="DataSourceName" value="jfkdatasource" />
34           <add key="IndexName" value="jfkindex" />
35           <add key="SkillsetName" value="jfkskillset" />
36           <add key="IndexerName" value="jfkindexer" />
37           <add key="SynonymMapName" value="cryptonyms" />
38           <add key="BlobContainerNameForImageStore" value="imagestoreblob" />
39       </appSettings>
40       <runtime>
```

Figure 4-16. *Pasting the values in the app config file*

You will need to get the SearchServiceQueryKey value from the deployed section. For the deployment, click the **Overview** tab, as shown in Figure 4-17.

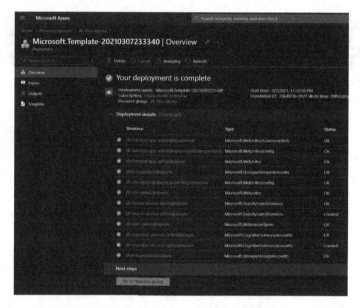

Figure 4-17. *Getting the SearchServiceQueryKey value*

Select the deployed service (as shown in Figure 4-18), and then click
Operation details on the right.

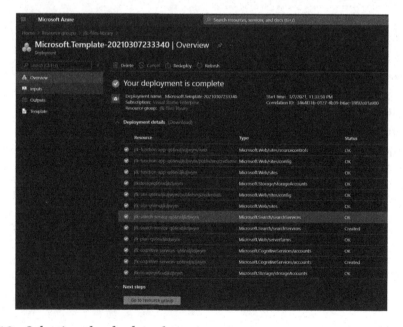

Figure 4-18. *Selecting the deployed service*

From the details screen, click **Keys** in the left menu, under Settings (as shown in Figure 4-19). Copy the SearchServiceQueryKey value, and then paste it in the application configuration file.

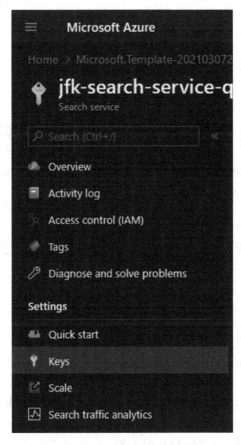

Figure 4-19. *The Keys option*

Now build the API solution, which will create the application keys (as shown in Figure 4-20). You will come back to this window after you run the npm steps in the following.

```
Deleting Data Source, Index, Indexer, Skillset and SynonymMap if they exist...
Creating Blob Container for Image Store Skill...
Creating Data Source...
Creating Skill Set...
Creating Synonym Map...
Creating Index...
Creating Indexer...
Setting Website Keys...
Website keys have been set.  Please build the website and then return here and press any key to continue.
```

Figure 4-20. *The setting keys in the API solution*

4.   The next step is to build the web application, which is built with Node.js. Open up the Node.js command prompt, and navigate to the front-end folder. See Figure 4-21.

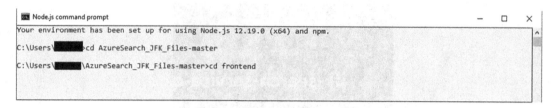

Figure 4-21. *Changing the directory to the downloaded folder*

Run the **npm install** command to build the application (as shown in Figure 4-22). It will get all the dependencies.

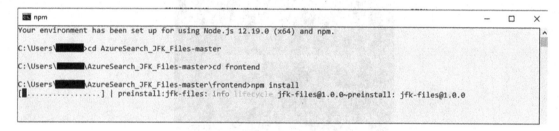

Figure 4-22. *Running the command, npm install*

Once the npm install is completed, run the **npm run build:prod** command to run the web application, which will build and deploy the web application. See Figure 4-23.

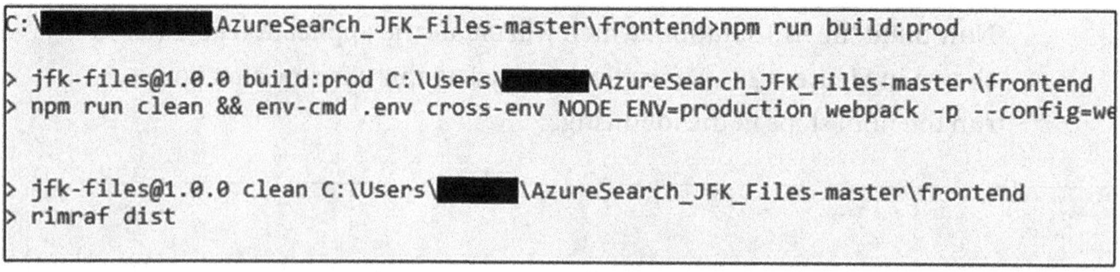

```
C:\          AzureSearch_JFK_Files-master\frontend>npm run build:prod

> jfk-files@1.0.0 build:prod C:\Users\    \AzureSearch_JFK_Files-master\frontend
> npm run clean && env-cmd .env cross-env NODE_ENV=production webpack -p --config=we

> jfk-files@1.0.0 clean C:\Users\    \AzureSearch_JFK_Files-master\frontend
> rimraf dist
```

Figure 4-23. *Running the command, npm run build:prod*

Next, return to the previous screen from step 3 (the Overview), and then press any key to complete the build (as shown in Figure 4-24). Then navigate to the URL in the console, to open the web application.

Figure 4-24. *Returning to the previous screen to complete the build*

The build and deployment are now complete, and the associated JFK Files application can be accessed, as shown in Figure 4-25.

Figure 4-25. *Searching in the JFK Files app*

Here, you can search for a term, like "security," and see the associated
results from the images of handwritten notes, as well as the typed
PDF files. It shows the associated entities on the left pane and the
documents containing the terms in the main window (as shown in
Figure 4-26).

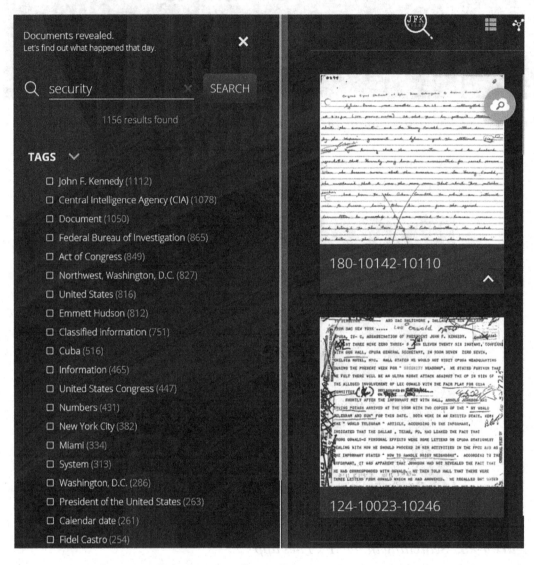

Figure 4-26. *The JFK Files app search results*

This concludes the installation, configuration, and running the JFK Files application. It is very impressive to see how 34,000 pages related to the JFK assassination were catalogued and interpreted, using Azure Cognitive Services, Azure Search, Computer Vision API (which uses text recognition APIs for OCR), Text Analytics API, and Custom Skills. These features were used to build a complex cognitive search solution. It can even correlate and search on CIA's cryptonyms, such as GPFLoor to Oswald[1].

JFK Files is not a hypothetical exercise in futility, but a realistic solution for enterprise data that is largely unstructured, that exists in multiple formats, and that comes from all types of data sources. By applying and modifying the natural language processing pipeline prescribed in the JFK solution, you can easily use it for a variety of enterprise solutions, where similar conditions are met. In the next example, we will explore the transformer language models.

The Muppet Models – Transformers for NLP

No discussion in natural language processing, especially around unstructured text and models, is complete without the mention of BERT, ELmo, Grover, RoBERTa, ERNIEs, and a KERMIT. It is not because the ACL (Association for Computational Linguistics) flagship conference was somehow co-located with a Sesame Street convention (where Big Bird definitely did not give a memorable keynote speech), but it's because when language model papers introduced Sesame Street-related acronyms, the inside joke kind of got out of hand (and many people started contributing to the bad joke).

An overview of transformer models and their intricate details would require a chapter of its own, but in brevity, the table in Figure 4-27 shows an overview of some of the popular Muppet models.

---

[1]AI enrichment with image and natural language processing in Azure Cognitive Search
https://docs.microsoft.com/en-us/azure/architecture/solution-ideas/articles/
cognitive-search-with-skillsets

	BERT	RoBERTa	DistilBERT	XLNet
Size (millions)	**Base:** 110 **Large:** 340	**Base:** 110 **Large:** 340	**Base:** 66	**Base:** ~110 **Large:** ~340
Training Time	**Base:** 8 x V100 x 12 days* **Large:** 64 TPU Chips x 4 days (or 280 x V100 x 1 days*)	**Large:** 1024 x V100 x 1 day; 4-5 times more than BERT.	**Base:** 8 x V100 x 3.5 days; 4 times less than BERT.	**Large:** 512 TPU Chips x 2.5 days; 5 times more than BERT.
Performance	Outperforms state-of-the-art in Oct 2018	2-20% improvement over BERT	3% degradation from BERT	2-15% improvement over BERT
Data	16 GB BERT data (Books Corpus + Wikipedia). 3.3 Billion words.	160 GB (16 GB BERT data + 144 GB additional)	16 GB BERT data. 3.3 Billion words.	**Base:** 16 GB BERT data **Large:** 113 GB (16 GB BERT data + 97 GB additional). 33 Billion words.
Method	BERT (Bidirectional Transformer with MLM and NSP)	BERT without NSP**	BERT Distillation	Bidirectional Transformer with Permutation based modeling

Figure 4-27. *A table with some popular Muppet models*

Named Entity Recognition with Fine-Tuned BERT

Named entity recognition (NER) is one of the key capabilities of any natural language processing system, and therefore it's among the fundamental features offered by language models. It is the capability of extracting entities from unstructured datasets, such as geographical entities (city, country, and so on), organizational entity (the name of an organization), personal entity (names and identifiers associated with people), geopolitical entity (countries, organizations), time, event, natural phenomenon, seasonality (holidays), custom entities, and so on.

In this example, we will use PyTorch Pretrained BERT on Azure ML, for named entity recognition. We will use Google Colaboratory, a free tool for running Jupyter notebooks. (It's often called Google Colab. The name is a pun that mixes the words *laboratory* and *collaboration*.) However, you can also complete the same exercise on your local machine via Anaconda or on Azure Machine Learning via Azure notebooks.

Use the following steps to complete this example.

1. Open the Jupyter notebook in Colab from `https://github.com/microsoft/AzureML-BERT/blob/master/finetune/PyTorch/notebooks/Pretrained-BERT-NER.ipynb` (as shown in Figure 4-28).

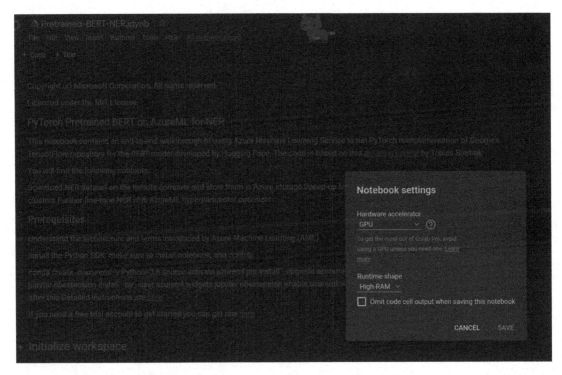

Figure 4-28. *Opening the BERT NER colab with Azure ML*

Populate your subscription ID and resource group information. You would also need to perform the interactive authentication by opening the browser and entering the provided code. See Figure 4-29.

```
1 from azureml.core import Workspace
2
3 subscription_id = '                              '
4 resource_group  = 'demo-resource-group'
5 workspace_name  = 'adnan-demo-workspace'
6 ws = Workspace(subscription_id = subscription_id, resource_group = resource_group, workspace_name = workspace_name)
7 ws.write_config()
8
9 try:
10     ws = Workspace.from_config()
11     print(ws.name, ws.location, ws.resource_group, ws.location, sep='\t')
12     print('Library configuration succeeded')
13 except:
14     print('Workspace not found')

Performing interactive authentication. Please follow the instructions on the terminal.
To sign in, use a web browser to open the page https://microsoft.com/devicelogin and enter the code E8LB8GMGB to authenticate.
```

Figure 4-29. *Populating your information*

You now need to initialize the workspace and install the Azure ML SDK for notebooks (as shown in Figure 4-30).

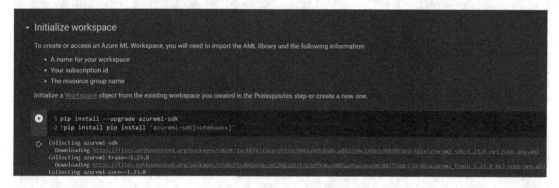

Figure 4-30. *Installing the Azure ML SDK for notebooks*

The next step is to create (or use) a compute cluster target for execution. In this case, we chose a standard single GPU NC_6 VM, while our region is set to western United States. See Figure 4-31.

```python
1 from azureml.core.compute import ComputeTarget, AmlCompute
2 from azureml.core.compute_target import ComputeTargetException
3
4 # Choose a name for your CPU cluster
5 cluster_name = "cluster"
6
7 # Verify that cluster does not exist already
8 try:
9     cluster = ComputeTarget(workspace=ws, name=cluster_name)
10     print('Found existing cluster, use it.')
11 except ComputeTargetException:
12     compute_config = AmlCompute.provisioning_configuration(vm_size='Standard_NC6',
13                                                            min_nodes=1,
14                                                            max_nodes=4)
15     cluster = ComputeTarget.create(ws, cluster_name, compute_config)
16
17 cluster.wait_for_completion(show_output=True)
```

Figure 4-31. *Creating a compute cluster*

However, we ran into an error! You learn more from your mistakes than if everything works smoothly. In this case, NC6 type VM is not available in the region. You will see the error message, "STANDARD_NC6 is not supported in region westus. Please choose a different VM size." (See Figure 4-32.)

It is important to note that all the compute types are not available in all the regions, and therefore, it's important to choose the right region. You can read more about regions and VMs at `https://docs.microsoft.com/ azure/virtual-machines/regions` (on Microsoft Docs).

```
 7 # Verify that cluster does not exist already
 8 try:
 9     cluster = ComputeTarget(workspace=ws, name=cluster_name)
10     print('Found existing cluster, use it.')
11 except ComputeTargetException:
12     compute_config = AmlCompute.provisioning_configuration(vm_size='Standard_NC6',
13                                                            min_nodes=1,
14                                                            max_nodes=4)
15     cluster = ComputeTarget.create(ws, cluster_name, compute_config)
16
17 cluster.wait_for_completion(show_output=True)
```

```
Creating.
FailedProvisioning operation finished, operation "Failed"
-------------------------------------------------------------------------
ComputeTargetException                   Traceback (most recent call last)
<ipython-input-4-8bed41ee17e4> in <module>()
     15         cluster = ComputeTarget.create(ws, cluster_name, compute_config)
     16
---> 17 cluster.wait_for_completion(show_output=True)

                       ↕ 2 frames
/usr/local/lib/python3.7/dist-packages/azureml/core/compute/compute.py in wait_for_completion(self,
show_output, is_delete_operation)
    566                         'state, current provisioning state: {}\n'
    567                         'Provisioning operation error:\n'
--> 568                         '{}'.format(self.provisioning_state,
error_response))
    569             except ComputeTargetException as e:
    570                 if e.message == 'No operation endpoint':

ComputeTargetException: ComputeTargetException:
        Message: Compute object provisioning polling reached non-successful terminal state, current
provisioning state: Failed
Provisioning operation error:
{'code': 'VmSizeNotSupported', 'message': 'STANDARD_NC6 is not supported in region westus. Please
choose a different VM size.', 'details': []}
        InnerException None
        ErrorResponse
{
    "error": {
        "message": "Compute object provisioning polling reached non-successful terminal state, current
provisioning state: Failed\nProvisioning operation error:\n{'code': 'VmSizeNotSupported', 'message':
'STANDARD_NC6 is not supported in region westus. Please choose a different VM size.', 'details': []}"
    }
}
```

Figure 4-32. *ComputeTarget creation error*

To fix this issue, go to the Azure console, and change the region to the eastern United States by creating a new workspace in eastern United States, where the VM is available. Now you can successfully create the compute target, as shown in Figure 4-33.

```
1 from azureml.core.compute import ComputeTarget, AmlCompute
2 from azureml.core.compute_target import ComputeTargetException
3
4 # Choose a name for your CPU cluster
5 cluster_name = "cluster"
6
7 # Verify that cluster does not exist already
8 try:
9     cluster = ComputeTarget(workspace=ws, name=cluster_name)
10     print('Found existing cluster, use it.')
11 except ComputeTargetException:
12     compute_config = AmlCompute.provisioning_configuration(vm_size='Standard_NC6',
13                                                            min_nodes=1,
14                                                            max_nodes=4)
15     cluster = ComputeTarget.create(ws, cluster_name, compute_config)
16
17 cluster.wait_for_completion(show_output=True)
```

```
Creating...
SucceededProvisioning operation finished, operation "Succeeded"
Succeeded.....................
AmlCompute wait for completion finished

Minimum number of nodes requested have been provisioned
```

Figure 4-33. ComputeTarget creation successful

For the compute target setup, let's upload the NER dataset. In this example, we will use the annotated corpus using GMB (Groningen Meaning Bank) for entity classification from Kaggle[2]. First, download the NER dataset file from Kaggle, and then upload it to the notebook. Then, use the commands shown in Figure 4-34, to upload the files to the Azure ML workspace blob store.

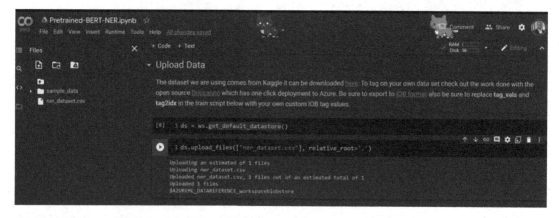

Figure 4-34. Uploading the dataset

The next step is to use this NER dataset and fine-tune the model. There are a few data engineering steps you'll need to take, such as replacing the tag values and tokenization. You'll use the BertForTokenClassification class for token-level predictions, as a fine-tuning model, which wraps the actual BertModel and adds token-level classifiers. The notebook creates the trian.py file, which is used to train the model (as shown in Figure 4-35).

```python
1
2  import argparse
3  import os
4  import pandas as pd
5  import numpy as np
6  from tqdm import tqdm, trange
7
8  import torch
9  from torch.optim import Adam
10 from torch.utils.data import TensorDataset, DataLoader, RandomSampler, SequentialSampler
11 from keras.preprocessing.sequence import pad_sequences
12 from sklearn.model_selection import train_test_split
13 from pytorch_pretrained_bert import BertTokenizer, BertConfig
14 from pytorch_pretrained_bert import BertForTokenClassification, BertAdam
15
16 from seqeval.metrics import f1_score
17
18
19 from azureml.core import Run
20
21 class SentenceGetter(object):
22
23     def __init__(self, data):
24         self.n_sent = 1
25         self.data = data
26         self.empty = False
27         agg_func = lambda s: [(w, p, t) for w, p, t in zip(s["Word"].values.tolist(),
28                                                            s["POS"].values.tolist(),
29                                                            s["Tag"].values.tolist())]
30         self.grouped = self.data.groupby("Sentence #").apply(agg_func)
31         self.sentences = [s for s in self.grouped]
32
33     def get_next(self):
34         try:
35             s = self.grouped["Sentence: {}".format(self.n_sent)]
36             self.n_sent += 1
37             return s
38         except:
39             return None
```

Figure 4-35. *Training a Python file for the BERT fine-tuning*

Now that you have the training file, create an Azure ML experiment[3], using train.py as the parameter for an entry script (which is the script used to run the experiment). See Figure 4-36.

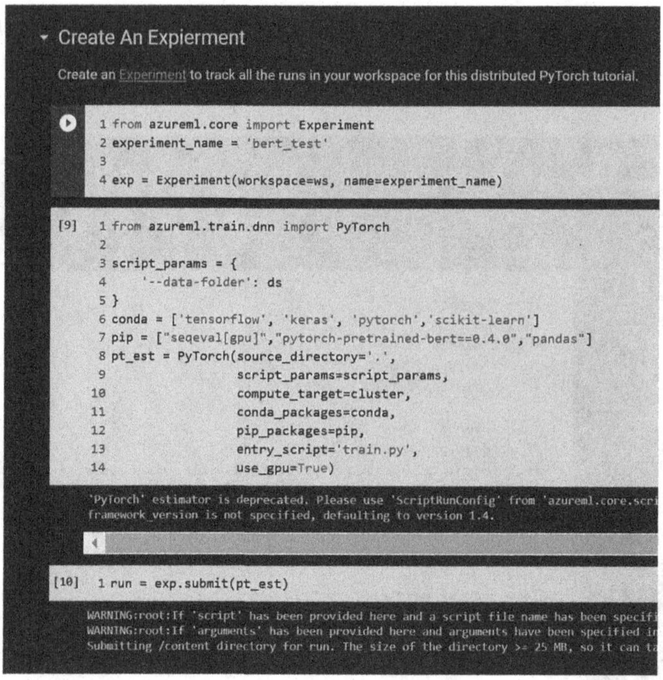

Figure 4-36.   *Creating the experiment to fine-tune the model*

You can track the running experiment in the Experiments tab, in the Azure Machine Learning console. See Figure 4-37.

---

[3]Create an experiment to track all the runs https://docs.microsoft.com/azure/machine-learning/service/concept-azure-machine-learning-architecture#experiment/?WT.mc_id=bert-notebook-abornst

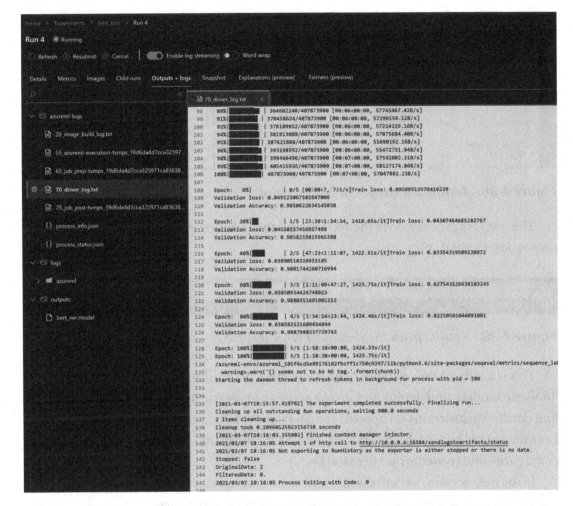

Figure 4-37. *Training experiment running, to fine-tune the model*

Once the experiment is completed, the service can be deployed in ACI and then invoked. Now you can see the named entity recognition results from this new fine-tuned model (as shown in Figure 4-38).

```
Test Deployed NER Service

 1 import requests
 2 import json
 3
 4 # send a random row from the test set to score
 5 input_data = "Microsoft's new open-source tool could stop your AI from getting hacked."
 6
 7 headers = {'Content-Type':'application/json'}
 8
 9 resp = requests.post(service.scoring_uri, input_data, headers=headers)
10 print("prediction:", " ".join([f"{t} ({l})" for t, l in zip(input_data.split(), json.loads(resp.text))]))
```

***Figure 4-38.** Testing the deployed NER service*

The text gets processed via the service, and the resulting entities are identified and shown. See Figure 4-39.

```
1 print ("prediction: Microsoft's (B-org) new (O) open-source (B-org) tool (O) could (O) stop (O) your (O) AI (I-org) from (O)  getting (O) hacked.")
prediction: Microsoft's (B-org) new (O) open-source (B-org) tool (O) could (O) stop (O) your (O) AI (I-org) from (O)  getting (O) hacked.
```

***Figure 4-39.** Results from the deployed NER service*

This is a rather brief overview of how you can use transformer models in Azure ML, fine-tune models, and deploy and invoke services to get the desired results. You can find detailed instructions on how to set up a Python environment (for Azure Machine Learning) at https://docs.microsoft.com/azure/machine-learning/how-to-configure-environment (on Microsoft Docs).

In the next section, we will discuss how unstructured text and models are used across the ecosystem, taking a deeper look at some related use cases.

Cognitive Services and language models are ubiquitously used across the application ecosystem, to build intelligent applications. Microsoft Learning Tools (also referred to as Digital Learning Tools) leverage these capabilities, for the education. For example, Microsoft Immersive Reader[4] utilizes the Azure Text Analytics capabilities, including comprehensive natural language analysis (for sentiment, relations, trends, and so on).

The Text Analytics service is described as an *"AI service that uncovers insights such as sentiment analysis, entities, relations, and key phrases in unstructured text."* It provides capabilities, such as broad entity extraction, sentiment analysis, language detection, and flexible deployment. See Figure 4-40 for the Text Analytics service homepage.

---

[4]www.microsoft.com/en-us/education/products/learning-tools

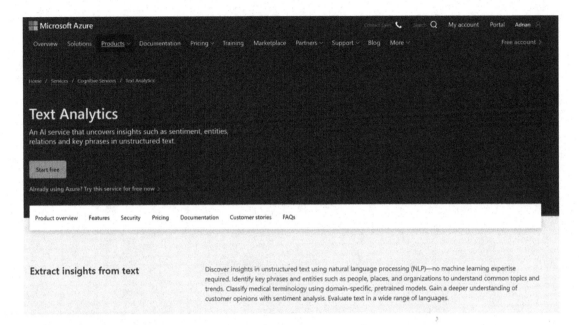

Figure 4-40. *Text Analytics service homepage*

The Text Analytics service is available in the cloud, on premises, and as an edge deployment, which makes it a great utility for a variety of different enterprise needs. For instance, you can use it on premises as a container. Details on installing and running Text Analytics containers can be found on Microsoft Docs[5].

The open source alternative to this service is achieved with custom-trained models, with libraries like spaCy, NLTK, and CoreNLP, or with cloud providers like Hugging Face, OpenAI, and so on. However, using the Text Analytics service separates out the implementation details, and it helps you work in abstraction. Instead of fine-tuning custom models to identify and categorize important concepts, you can utilize the wide range of prebuilt entities and hundreds of personally identifiable information (PII) data elements, including protected health information (PHI). You can extract relevant phrases, topic models, customer sentiment analysis, and medical data (Text Analytics for health is in preview[6]).

---

[5]https://docs.microsoft.com/en-us/azure/cognitive-services/text-analytics/how-tos/
text-analytics-how-to-install-containers?tabs=sentiment

[6]https://docs.microsoft.com/en-us/azure/cognitive-services/text-analytics/how-tos/
text-analytics-for-health?tabs=ner

Summary of the Language API

In this chapter, we explored some of the key concepts around language APIs and their use for processing unstructured text and models. We demonstrated, with examples, how Azure Cognitive Services and Azure Machine Learning can be used to build a comprehensive search engine. We also studied transformer models and how to fine-tune one, using Azure Machine Learning.

In the next chapter, we will continue these topics, by exploring speech and speech services as the next step in your Cognitive Services journey.

CHAPTER 5

Speech – Talk to Your Application

In the previous chapter, we saw how text models can be used to build enriched text and to build smart search applications from unstructured data. In this chapter, we will add a capability to our cognitive computing toolkit – how to understand the request from a user and process these requests via speech. To extend this skillset, you will see how you can take user input via speech, create custom input events, and interact with your user.

Interaction, in communication, is one of the important behaviors of humans. With the help of communication, you can describe the things you see and use. Communication can be verbal or nonverbal, but the aim would be the same in either case. Your speech plays an important role to describe and commute your words. Your verbal communication is flawless, if your speech is delivered in such a way that your audience can understand it. In the same way, your nonverbal communication is mostly depending upon the text. Your text should be precise and understandable.

In a real-world scenario, there are a lot of hurdles for someone to overcome, in order to communicate everything they need to convey, perfectly, in nonverbal communication (text). Others can do the same, more easily, in verbal communication (speech).

The aim of this chapter is to provide insight on speech services by evaluating and translating text to speech and vice versa. We will also build a comprehensive sample that provides cognitive capabilities, such as speech recognition with microphone input, continuous recognition with file input, customized models, pull and push audio streams, keyword spotting, translation with microphone input, file input and audio streams, intent recognition, speech synthesis, keyword recognition, language detection, and much more. All these capabilities can be used in a variety of enterprise use cases.

© Ed Price, Adnan Masood, and Gaurav Aroraa 2021
E. Price et al., *Hands-on Azure Cognitive Services*, https://doi.org/10.1007/978-1-4842-7249-7_5

Throughout this chapter, the following topics will be covered:

- Understanding speech and speech services

- Speech to text – converting spoken audio to text for interaction

- Cognitive speech search with LUIS and Speech Studio

- Custom Keyword using Speech Studio

- Summary of the Speech API

Understanding Speech and Speech Services

As humans, we have special abilities, and one of those abilities is the ability to express our thoughts and feelings, to explain specific work and tasks by generating sounds. These sounds could be anything, vocal music, words, or a song. Anything classified as a linguistic sound can be translated to become understood in any language. This is what we call *speech*. The aim of speech is to connect or communicate with an audience, in order to deliver a specific message. The message may differ from one place to another place. (For example, in a classroom, a teacher's message is related to the subject for her or his students.) Similarly, when we talk about the application in software development, it must communicate directly with its users. This communication can be through voice (speech) or text.

Microsoft Azure Cognitive Services provides the facility to develop smart applications, with the help of APIs and the Azure Speech service.

Comprehensive Privacy and Security[1]

- The Speech service, part of Azure Cognitive Services, is certified by SOC, FedRAMP, PCI DSS, HIPAA, HITECH, and ISO.

- Your data remains yours. Your audio input and transcription data aren't logged during audio processing.

- View and delete your Custom Speech data and models at any time. Your data is encrypted while it's in storage.

---

[1]Azure Cognitive Services – speech to text. https://azure.microsoft.com/en-us/services/cognitive-services/speech-to-text/

- Backed by the Azure infrastructure, the Speech service offers enterprise-grade security, availability, compliance, and manageability.

Microsoft Azure has introduced the Speech service, which also replaced the Bing Speech API and Translator Speech.

Azure Cognitive Services provides the following tools to develop speech-enabled smart applications:

- Azure Speech CLI – `https://docs.microsoft.com/en-us/azure/cognitive-services/speech-service/spx-overview`

- Cognitive Services Speech Studio – `https://speech.microsoft.com/`

- Speech Devices SDK – `https://docs.microsoft.com/en-us/azure/cognitive-services/speech-service/speech-devices-sdk-quickstart`

- Speech service API – `https://docs.microsoft.com/en-us/azure/cognitive-services/speech-service/overview#reference-docs`

Getting Started

If you want to start or want to try the Speech service, you can try it with a free account by following these steps:

1. If you don't have an Azure account, you can create a free account: `https://azure.microsoft.com/en-in/free/`.

2. Sign into the Azure portal: `https://portal.azure.com/`.

3. From the search text box, search for "Speech."

4. Select "Speech" services, and then click **Create** to create a new resource (as shown in Figure 5-1).

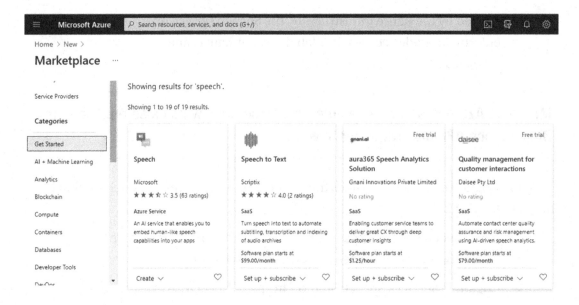

Figure 5-1. *Creating a new resource for the Speech service*

Translating Speech Real Time into Your Application

Azure Cognitive Services are backed by a world-class model deployment technology, which was built by top experts. There are many plans and offers for you to use a pay-as-you-go model, instead of investing in development and infrastructure that you may need if you choose to develop and host your models.

Azure Marketplace is a one-stop place to get all the services. You will be sent to Azure Marketplace, as soon as you click **Add** or **Create cognitive services** (as shown in Figure 5-2).

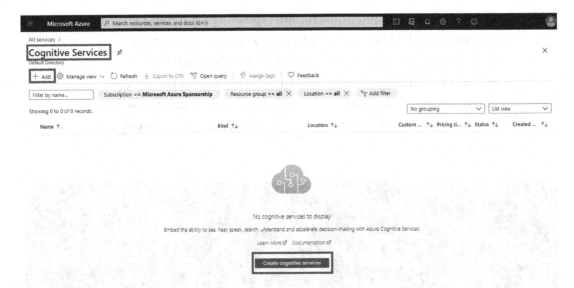

Figure 5-2. *Add/Create cognitive services*

Before getting started with the Azure Speech service, let's take a look at some important considerations. Due to frequent updates, new regions, and languages being added, your environment will get out of sync as soon as this book gets published. Therefore, we have included the following links that will provide the most up-to-date information:

- Language, voice support – https://docs.microsoft.com/en-us/azure/cognitive-services/speech-service/language-support

- Service availability, region wise – https://docs.microsoft.com/en-us/azure/cognitive-services/speech-service/regions

- Pricing – https://azure.microsoft.com/en-us/pricing/details/cognitive-services/speech-services/

Speech to Text – Converting Spoken Audio to Text for Interaction

In this first example, we will see how to transcribe conversations in C# by using .NET Framework for Windows. Follow these steps:

Clone the conversation transcription repository from the cognitive
services speech SDK repository (`https://github.com/Azure-`
`Samples/cognitive-services-speech-sdk/tree/master/`
`quickstart/csharp/dotnet/conversation-transcription`). See
Figure 5-3. Then navigate to the equivalent folder (cognitive-services-
speech-sdk/quickstart/csharp/dotnet/**conversation-transcription**/)
in your system.

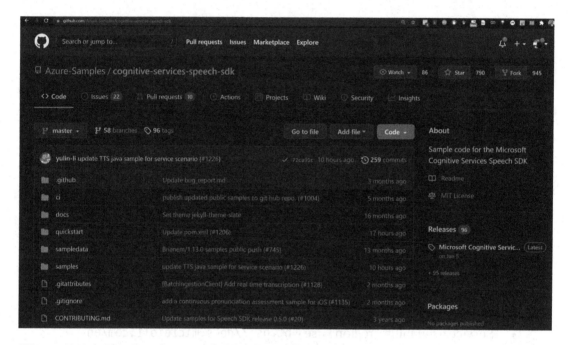

Figure 5-3. *Cloning the speech SDK GitHub repository*

1. Open up Visual Studio, and then open the solution file, from
   the cloned path (quickstart/csharp/dotnet/conversation-
   transcription). See Figure 5-4.

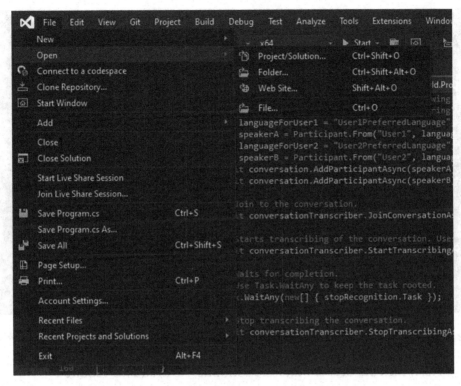

Figure 5-4. *Visual Studio – opening the solution from the path*

    2.   Open the Azure services console, and then click **Create a resource** (as shown in Figure 5-5).

Figure 5-5. *Azure services – Create a resource*

Select **Speech** from the marketplace, and then click the **Speech** resource by Microsoft. See Figure 5-6.

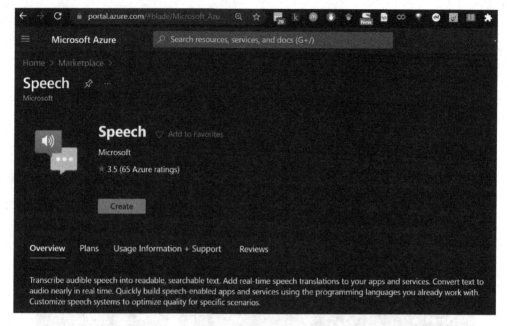

Figure 5-6. *Azure services – creating a speech resource from the marketplace*

Next, click the **Create** button to continue. Next, add the required
service parameters, such as the name, subscription, location, pricing
tier, and resource group (as shown in Figure 5-7).

Figure 5-7. *Azure services – providing the required information*

Click **Create** to continue. This deploys the service. You will see the notification of deployment as it starts, as shown in Figure 5-8.

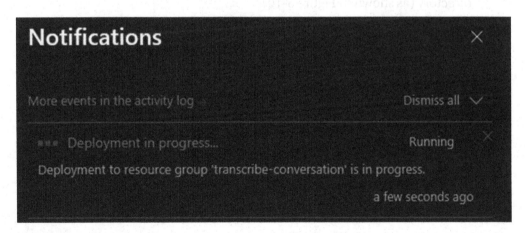

Figure 5-8. *Azure services – speech deployment in progress*

Once the deployment completes, you will see the keys and the respective endpoints, as shown in Figure 5-9.

Figure 5-9. *Azure Speech service – keys and endpoint*

3. You can copy the key and service region (location) from this page. Then, use it to replace the tags in the solution file of the open directory (as shown in Figure 5-10).

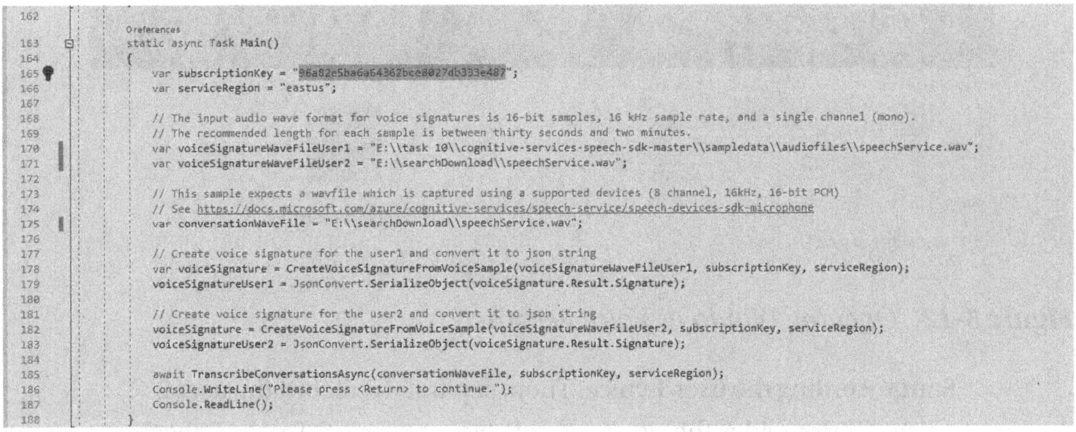

```
162
163      static async Task Main()
164      {
165          var subscriptionKey = "96a82e5ba6a64367bce8027db333e487";
166          var serviceRegion = "eastus";
167
168          // The input audio wave format for voice signatures is 16-bit samples, 16 kHz sample rate, and a single channel (mono).
169          // The recommended length for each sample is between thirty seconds and two minutes.
170          var voiceSignatureWaveFileUser1 = "E:\\task-10\\cognitive-services-speech-sdk-master\\sampledata\\audiofiles\\speechService.wav";
171          var voiceSignatureWaveFileUser2 = "E:\\searchDownload\\speechService.wav";
172
173          // This sample expects a wavfile which is captured using a supported devices (8 channel, 16kHz, 16-bit PCM)
174          // See https://docs.microsoft.com/azure/cognitive-services/speech-service/speech-devices-sdk-microphone
175          var conversationWaveFile = "E:\\searchDownload\\speechService.wav";
176
177          // Create voice signature for the user1 and convert it to json string
178          var voiceSignature = CreateVoiceSignatureFromVoiceSample(voiceSignatureWaveFileUser1, subscriptionKey, serviceRegion);
179          voiceSignatureUser1 = JsonConvert.SerializeObject(voiceSignature.Result.Signature);
180
181          // Create voice signature for the user2 and convert it to json string
182          voiceSignature = CreateVoiceSignatureFromVoiceSample(voiceSignatureWaveFileUser2, subscriptionKey, serviceRegion);
183          voiceSignatureUser2 = JsonConvert.SerializeObject(voiceSignature.Result.Signature);
184
185          await TranscribeConversationsAsync(conversationWaveFile, subscriptionKey, serviceRegion);
186          Console.WriteLine("Please press <Return> to continue.");
187          Console.ReadLine();
188      }
```

Figure 5-10. *Azure Speech service – populate the keys and service region*

Edit the Program.cs source file, and then replace the YourSubscriptionKey and YourServiceRegion values, preferred language, and add a WAV file sample for a voice signature. See Figure 5-11.

```
170      var voiceSignatureWaveFileUser1 = "E:\\searchDownload\\user1.wav";
171      var voiceSignatureWaveFileUser2 = "E:\\searchDownload\\user2.wav";
172
173      // This sample expects a wavfile which is captured using a supported devices (8 channel, 16kHz, 16-bit PCM)
174      // See https://docs.microsoft.com/azure/cognitive-services/speech-service/speech-devices-sdk-microphone
175      var conversationWaveFile = "E:\\searchDownload\\speechService.wav";
176
```

Figure 5-11. *Azure Speech service – populating the keys and service region*

Using the key directly in code is not recommended. A secure way is to use key vault or use system variables.

This voice signature sample is for the voice being detected. The recommended length for each sample is between thirty seconds and two minutes. Cognitive Services expects a .wav file, which should be captured using a supported device (8 channel, 16kHz, 16-bit PCM). To create this WAV file, open up the Windows Voice Recorder or an audio recorder of your choice. Here, we are using the Windows Voice Recorder, as shown in Figure 5-12.

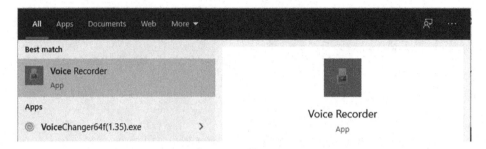

Figure 5-12. *Opening Windows Voice Recorder*

Start recording the user 1 voice. Then, separately, record the user 2 voice. This provides the signatures of your user voices for the Speech service. See Figure 5-13.

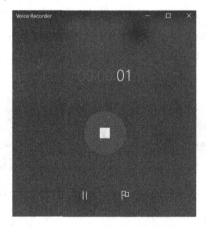

Figure 5-13. *Recording audio files of conversations*

You can also record a single conversation file, between user 1 and user 2. See Figure 5-14 for our three recorded audio files.

Figure 5-14. *The recorded audio files*

At this point, we have three .wav files. If you created the voices by using the given specifications[2], then place the file path in the program and proceed with the build. Otherwise, follow these steps to get the files in the right format.

4. Open Audacity[3] and navigate to the .wav file that you recorded. See Figure 5-15. Audacity is a free, open source, cross-platform audio software, which can be used to convert audio files into multiple formats.

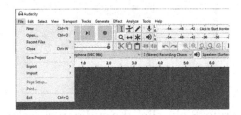

Figure 5-15. *Opening a file in Audacity*

As you can see in Figure 5-16 (via the red arrow), our recorded file does not meet the required specifications. Therefore, we need to modify the format.

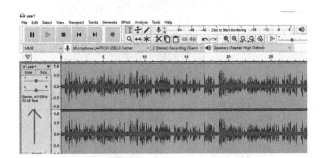

Figure 5-16. *Format details in Audacity*

---

[2]https://docs.microsoft.com/en-us/azure/cognitive-services/speech-service/ speech-devices-sdk-microphone

[3]www.audacityteam.org/

To edit the file format, right-click the track, select **Format**, and then change it to **16-bit PCM** (as shown in Figure 5-17).

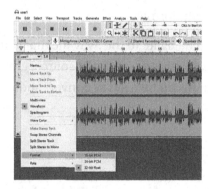

Figure 5-17. *Changing the format in Audacity*

Also, change the rate. Right-click the audio track, select **Rate**, and then click **16000 Hz** (as shown in Figure 5-18).

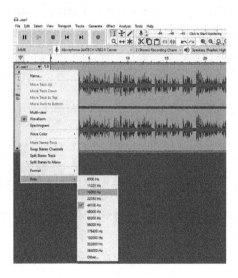

Figure 5-18. *Changing the rate in Audacity*

Next, to convert the file from stereo to mono, in the top file menu, click **Tracks**, select **Mix**, and then click **Mix Stereo Down to Mono**. See Figure 5-19.

Figure 5-19. *Changing the mix in Audacity*

At this point, the file will look like Figure 5-20.

Figure 5-20. *The file format changes in Audacity*

Next, export the file as a WAV format, and then save it to the directory of your choice. Click **File**, **Export**, and then **Export as WAV** (as shown in Figure 5-21). After that, a new dialog box appears.

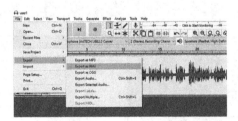

Figure 5-21. *Exporting as a WAV file*

Open the Advanced Mixing Options[4], and then drag the output channel slider to 8, as shown in Figure 5-22.

---

[4]Advanced Mixing Options https://manual.audacityteam.org/man/advanced_mixing_options.html#:~:text=The%20Advanced%20Mixing%20Options%20dialog,channels%20in%20the%20exported%20file.

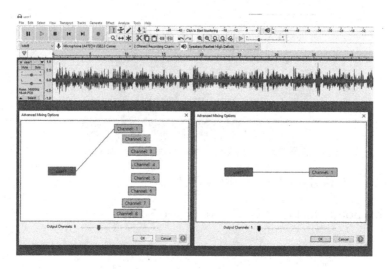

Figure 5-22. *Advanced Mixing Options*

Link all the channels to user 1, except for Channel 8 (as shown in Figure 5-23). Then click **OK**. Do the same step for the conversation file of your recorded files, to meet the specification.

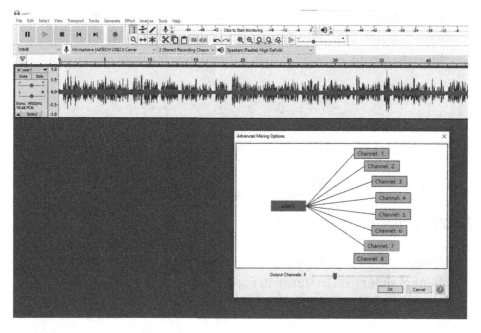

Figure 5-23. *Linking the channels*

Next, in the config file, add the file path accordingly, as shown in Figure 5-24.

Figure 5-24. *Modifying the config file*

Click the **Start** button to build and run the project (as shown in Figure 5-25).

```
E:\task 10\cognitive-services-speech-sdk-master\quickstart\csharp\dotnet\conversation-transcription\helloworld\bin\x64\Debug\helloworld.exe
TRANSCRIBING: Text=what is the speech services the speech service is the unification of speech attacks text to speech and speech translation
TRANSCRIBING: Text=what is the speech services the speech service is the unification of speech attacks text to speech and speech translation
TRANSCRIBING: Text=what is the speech services the speech service is the unification of speech attacks text to speech and speech translation
TRANSCRIBING: Text=what is the speech services the speech service is the unification of speech attacks text to speech and speech translation
TRANSCRIBING: Text=what is the speech services the speech service is the unification of speech attacks text to speech and speech translation
TRANSCRIBING: Text=what is the speech services the speech service is the unification of speech attacks text to speech and speech translation
TRANSCRIBING: Text=what is the speech services the speech service is the unification of speech attacks text to speech and speech translation
TRANSCRIBING: Text=what is the speech services the speech service is the unification of speech attacks text to speech and speech translation
TRANSCRIBED: Text= SpeakerId=$ref$
TRANSCRIBING: Text=what is the speech services the speech service is the unification of speech attacks text to speech and speech translation
TRANSCRIBING: Text=what is the speech services the speech service is the unification of speech attacks text to speech and speech translation
fied
TRANSCRIBING: Text=what is the speech services the speech service is the unification of speech attacks text to speech and speech translation
ified
```

Figure 5-25. *The program running and transcribing the text*

Upon a successful build and execution of the program, a console window will open up, showing the transcribed text as present in the audio file that you recorded.

There is a variety of practical applications that you can use for this transcription, from converting call center audio recordings into textual data for analytics to creating video subtitles. You might have already seen and used Microsoft Teams, Skype, and Zoom to create subtitles for taking notes, based on audio conversations. These transcriptions aren't perfect, but they are definitely getting better. Imagine performing a real-time escalation analysis, by combining the sentiment analysis of text APIs with the speech transcripts. Wouldn't that be an excellent way to analyze how an irate customer's call should be handled, in order to provide the best possible customer experience?

Now that you know how to create a speech solution in your own application, by using the Speech SDK and Azure Cognitive Services, what business problem would you want to solve? The possibilities are limitless.

Cognitive Speech Search with LUIS and Speech Studio

In the previous example, you learned how to transcribe an audio file. In this example, we will be using Custom Speech with the Microsoft LUIS (Language Understanding) service, to build a speech-enabled application. Follow these steps:

1. Create a speech resource from the marketplace, as shown in step 2 of the earlier example. We have done this a few times now, so hopefully you are well versed in it. Here, we would reuse the GitHub repository cloned in step 1 of the earlier example.

2. Start Visual Studio, and then select **File ➤ Open ➤ Project/ Solution**. Navigate to the folder that contains this sample, and then select the solution file contained within it. This should be the standard path (cognitive-services-speech-sdk/samples/csharp/dotnet-windows/console/).

3. Navigate to the intent_recognition_samples.cs, and then create an Azure Cognitive Services account (see Figure 5-26). LUIS is part of Cognitive Services and is dubbed as a *"machine learning-based service to build natural language into apps, bots, and IoT devices."* An enterprise ready and scalable service, LUIS provides a fast and efficient way to add language capabilities to your application. Upon logging into LUIS.AI, you will need to create a new Azure Cognitive Services account.

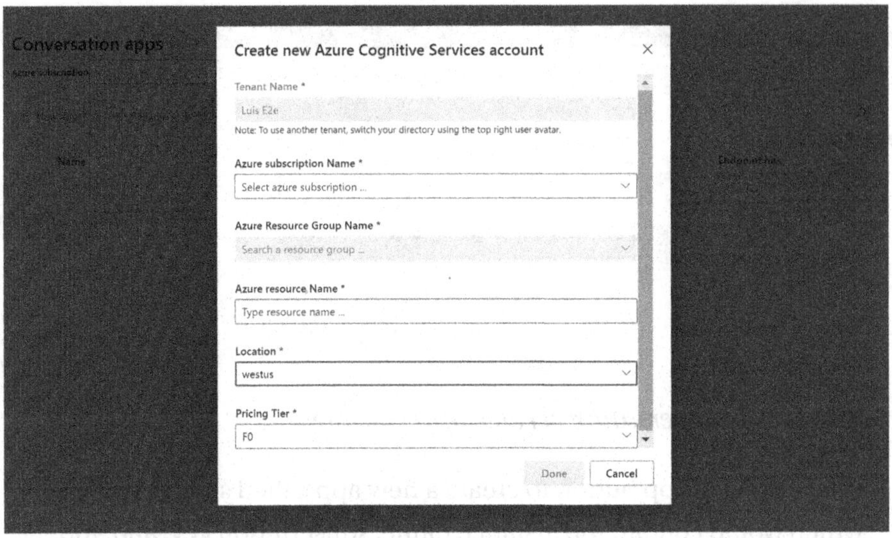

Figure 5-26. *Create new Azure Cognitive Services account*

Fill in the details to complete the process. At this point, you will see the notifications for your Azure account creation and respective resource creation, as shown in Figure 5-27.

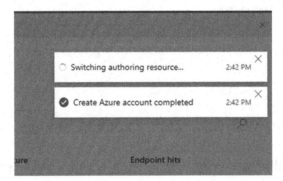

Figure 5-27. *The Azure Cognitive Services account notifications*

Once completed, proceed with the conversation apps authoring service console, as shown in Figure 5-28. Next, you will create a new app.

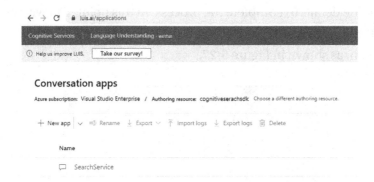

Figure 5-28. *LUIS conversation service app console*

Click the New App button to create a new app called search service
which would contain the instance name, subscription key, and app
ID. See Figure 5-29.

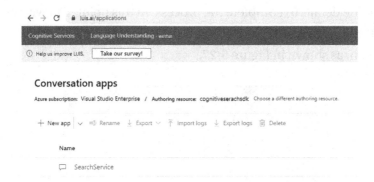

Figure 5-29. *LUIS conversation service – creating a new app*

Next, you will create a conversation app, called SearchService, as
shown earlier, in Figure 5-30. Open SearchService by double-clicking
it, and then create a new intent by clicking **Create new intent**. The
dialog box in Figure 5-30 opens. An *intent* is a task or action the user
mentions in the *utterance* to perform, for instance, order coffee,
check shipping status, etc. LUIS uses None intent as a fallback and
comes with pre-built domains which provide known intents with
utterances.

Figure 5-30. *LUIS conversation service – Create new intent*

Next, go to the created intent and add an example user input, as shown in Figure 5-31.

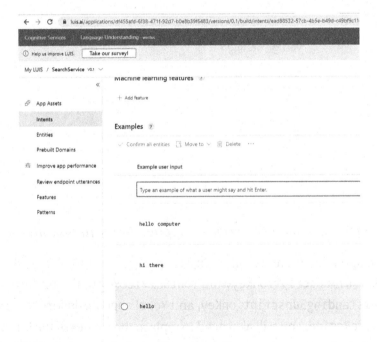

Figure 5-31. *LUIS conversation service – creating custom user input*

At this point, we have the keys for Language Understanding service (LUIS) to populate the solution file. You can click the manage tab, to see these keys (as shown in Figure 5-32).

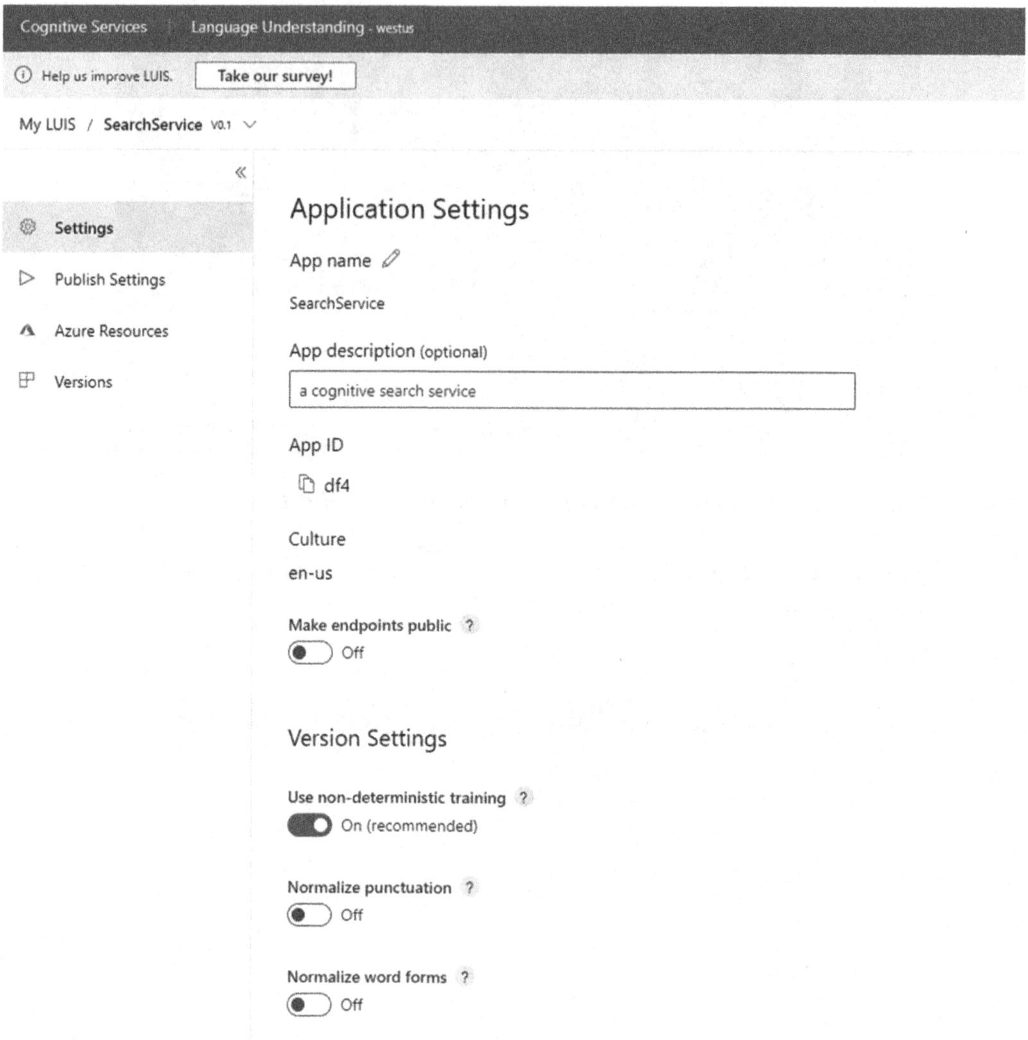

Figure 5-32. *LUIS conversation service – Application Settings and App ID*

Next, open the intent_recognition_samples.cs file, and replace the keys, YourSubscriptionKeywith, YourServiceRegion, YourLanguage UnderstandingSubscriptionKey, and YourLanguageUnderstanding ServiceRegion. You will also need to replace the names of the intents, such as YourLanguageUnderstandingIntentName1, YourLanguage UnderstandingIntentName2, and YourLanguageUnderstanding IntentName3, with the names of intents that LUIS recognizes. To find and replace these keys, in the top menu, click **Edit** and then **Find and Replace**. See Figure 5-33.

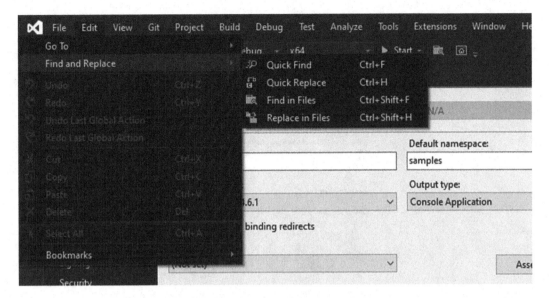

Figure 5-33. *Replacing the keys in the project in Visual Studio*

One last thing to consider is the associated model (such as YourKeywordRecognitionModelFile.table). You need to replace it with the location of your keyword recognition model file. In the next step, we will see how to obtain the table file from Speech Studio. Also, replace the YourKeyword trigger with the phrase from your keyword recognition model.

4. In order to obtain the YourKeywordRecognitionModelFile.table file, go to speech.microsoft.com. See Figure 5-34.

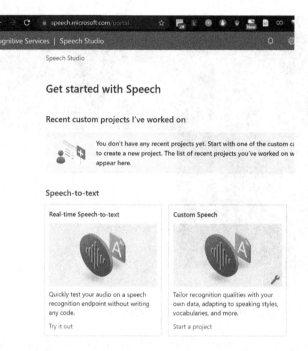

Figure 5-34. *The Speech Studio portal*

5. Log into Speech Studio, and then click **Custom Keyword**, as
    shown in Figure 5-35.

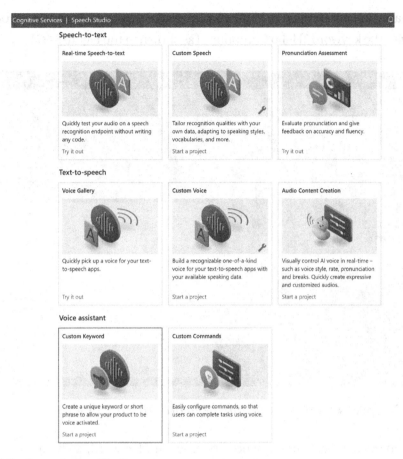

Figure 5-35. *Speech Studio features*

Custom Keyword is a voice assistant capability, to select keyword spotting. Click **New project** to create a new project (as shown in Figure 5-36).

Figure 5-36. *Speech Studio – Custom Keyword project*

151

Name the new project to **MyKeyboard**, and then populate the new model
with the keyword "Hello Computer" (as shown in the Figure 5-37).

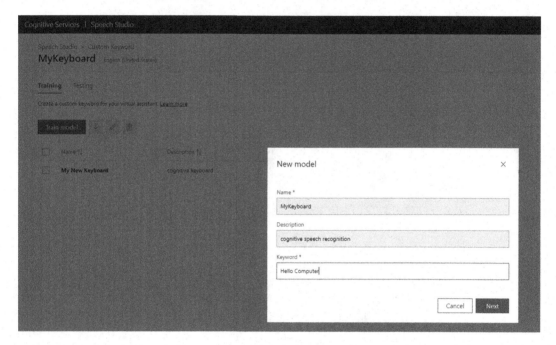

Figure 5-37. *Speech Studio – a custom keyword training model*

Add the details, to create a project. Add your keyword, with the
associated pronunciation. Listen to the different pronunciations of
the keyword, and then check the files that are the most suitable. Next,
click **Train**. See Figure 5-38.

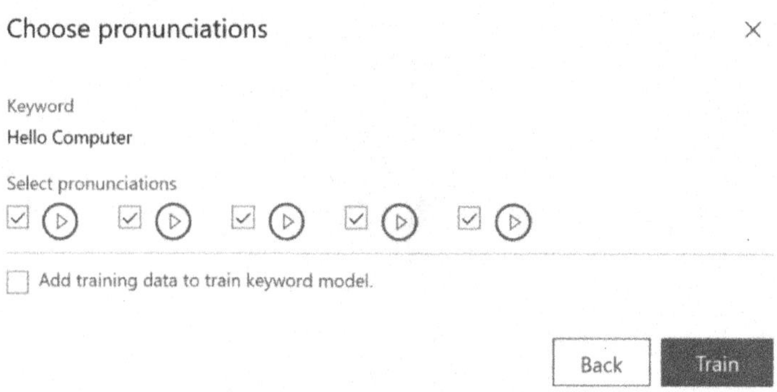

Figure 5-38. *Speech Studio training*

Once the training is completed, you can download the model table file (as shown in Figure 5-39).

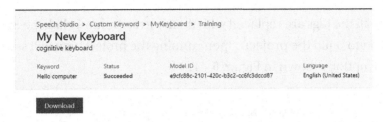

Figure 5-39. *Speech Studio – downloading the training file*

At this point, download the YourKeywordRecognitionModelFile. table file, and then replace that tag with the location of the downloaded file. See Figure 5-40.

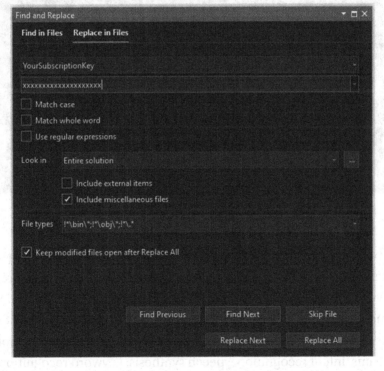

Figure 5-40. *Replacing the tags with the subscription values*

For the previously mentioned tags, simply use the find-and-replace option in Visual Studio. One by one, replace all the tags with their respective values.

After all the tags are replaced with their correct values, click the **Start** button to build the project. Upon running the project, you will see the list of options shown in Figure 5-41.

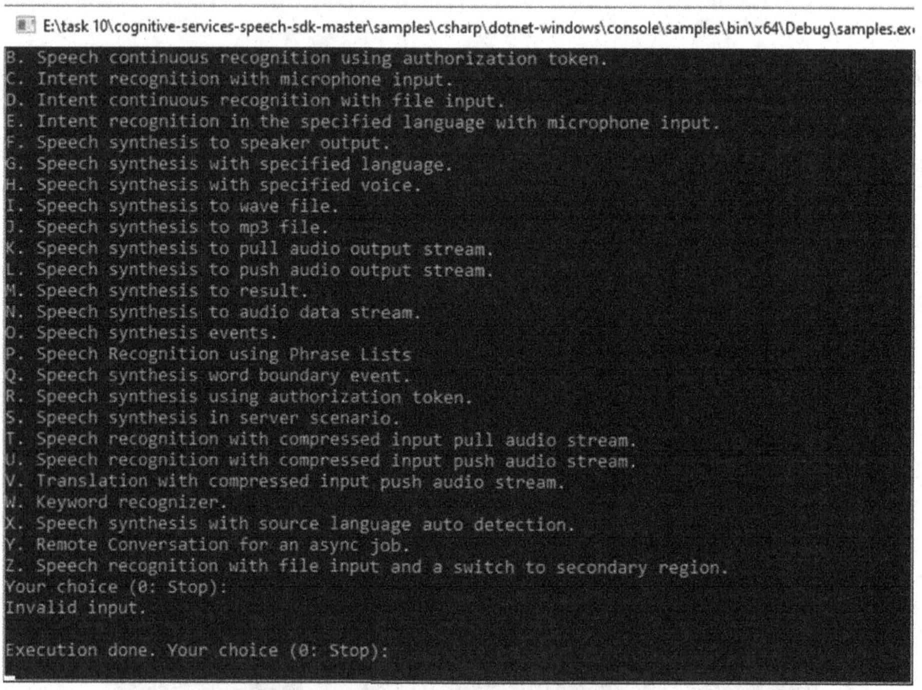

Figure 5-41. Running the speech application with LUIS

The final build will open a window with a different option to select and perform the respective operations, by using the Speech service and LUIS. This sample provides a comprehensive capability enumeration, such as speech recognition with microphone input, continuous recognition with file input, using a customized model, with pull and push audio streams, with keyword spotting, translation with microphone input, file input and audio streams, intent recognition, speech synthesis, keyword recognition, language detection, and much more.

Summary of the Speech API

In this chapter, we have worked with the Azure Cognitive Services Speech APIs. We have reviewed a few approaches of using the Speech service to help understand and interact with users by using speech, thus allowing your application to talk to your users.

In the next chapter, we will learn how to make applications smart enough so that they can make their own decisions. It's a scary new world. Stay tuned.

CHAPTER 6

Decision – Make Smarter Decisions in Your Applications

Decisions, decisions, decisions – sriracha or Tabasco, team Jacob or team Edward, PC or Mac, iPhone or Android, Fortnite or Apex, Coke or Pepsi, *Star Wars* or *Star Trek*... (Okay, that last one isn't that much of a discussion, but we digress.) Making decisions is possibly the most human thing we do. These decisions are based on inputs from sensory information available to us, coupled with some a priori observations, facts, our implicit biases, and lots of heuristics. This *cognitive computing* capability of being able to decide the next best action is a permeant fixture in decision sciences. Studying and capturing our innate skills of how we come to these decisions requires an understanding of how we ingest and process this information. Then we can conclude the results.

The nuances associated with decision making can be summed up quite well by, "I know it when I see it." This is a seminal quote by Justice Potter Stewart to describe his test for obscenity. Keeping it PG, in this chapter, we will see how APIs can be used for this purpose. This chapter will provide insight on decision services by adding content a moderation facility in the application. The chapter will also touch on a relatively new feature – Anomaly Detector. The following are some of the salient features of this chapter:

1. Understanding the decision service and Decision APIs

2. Creating an auto Content Moderator application

3. Creating personalized experiences with the Personalizer

4. Identifying future problems with the Anomaly Detector

© Ed Price, Adnan Masood, and Gaurav Aroraa 2021
E. Price et al., *Hands-on Azure Cognitive Services*, https://doi.org/10.1007/978-1-4842-7249-7_6

5. Exploring metrics with the Metrics Advisor service

6. Summary of the Decision API

So, without further ado, let's dive in!

Enterprise Decision Management is the art and science of using business rules, domain knowledge, predictive analytics, and big data to build and drive efficient business processes. It's an exciting topic, but nevertheless, it is beyond the scope of this manuscript.

Decision Services and APIs

Originally launched as "Project Custom Decision," it has now graduated to become part of the Personalizer service, joining the ranks of the Anomaly Detector, Content Moderator, and Metrics Advisor suite of Cognitive Services. This minor lesson in history is important since making a smarter decision requires a multifaceted skillset. Having one comprehensive "decision service" might not be the most effective way to address this nuanced discipline. Therefore, a combination of different skillsets is necessary, such as identifying outliers; finding out potentially offensive, racy, obscene, or unwanted contents; and creating tailored custom interactions for users.

Decision services are a critical part of the Azure Cognitive Services ecosystem, which also offers amazing capabilities in the areas of language, speech, vision, and web search (as you have seen in the earlier chapters). The current members of the decision service family include the following:

- **Anomaly Detector**

  Cognitive service to identify outliers and to pinpoint potential behaviors that are out of the ordinary

- **Content Moderator**

  Cognitive service that helps identify racy, obscene, and offensive contents across multiple modalities

- **Personalizer**

  Service to build a customized, individualized, and targeted experience that is curated for individual user preferences

- **Metrics Advisor (in preview at the time of this writing)**

  Anomaly detection core engine, customized for AIOps, such as detecting and diagnosing metrices for root cause analysis and finding issues in logs with time series analysis

In the rest of this chapter, we will review these services and build demo applications that will show you how to use them in your own projects.

Content Moderator Service

For today's digital economy, scale is the name of the game. The service that's offered must be able to scale to millions of users without these manual bottlenecks.

Extending on the "I know it when I see it" metaphor, an Internet scale system today cannot rely on human screenings for every video and image uploaded to the Internet. Platforms like YouTube get more than 500 hours of video contents uploaded every minute, which results in approximately 30,000 hours of hourly contents. The cost and resources required make it prohibitive for human screening. Human in the loop (HiTL) is the technique of keeping human interaction as part of the process. Ideally, the human interaction is limited only to the exceptional cases that really require human intervention for decision making. The rest is automated.

Content moderation is a required component in most modern websites that interact with their users. Some of the use cases include the following:

a. An organization offers a digital bank, which allows their customers to upload an image for their custom debit card. They want to ensure that the uploaded image does not contain any obscene or racy pictures.

b. For a company's intranet, a co-worker wants to upload the latest company holiday party photos. They want to ensure the images are safe for work.

    c. An organization's application moderates social media feeds, which are imported to be displayed on an employee's custom dashboard. It monitors Facebook, Twitter, and Instagram posts for contents about the company and associated social media hashtags, to ensure a strong reputation and full compliance.

    d. An organization builds a customer profile online, complete with photos and personalized information. They also want to ensure the pictures are not riddled with profane phrases.

    e. A company creates an ecommerce marketplace, where customers can make niche stores of their own and sell their products. They want to make sure that the pictures uploaded are suitable for the appropriate audience.

    f. A company provides a Discord-style chat room for their perspective customers. The customers can share live feedback about the company's products and services. The company wants to maintain decorum and keep the language moderated, to ensure no outwardly offensive terms are being used.

    g. A company creates a live-tour website, where users upload their TikTok adventures to their microsites, which promise authentic experiences. They want to make sure that these experiences, however authentic they might be, do not become a potential adult-content liability for their small business.

There are many more examples, but you get the point. Without sophisticated Cognitive Services and huge help from AI, a small business would be at a great disadvantage. Hypothetically speaking, large companies like YouTube, Twitter, Instagram, or Facebook might be able to hire an army of human reviewers to sift through every video, but you cannot afford to do that. However, that's okay because the Azure Content Moderator service is here to help. By the way, the companies we specified earlier heavily use AI and machine learning to perform content moderation, and they have humans in the loop as a secondary step. Some of them also have to deal with fake news; remember all those tweet warnings. Unfortunately, fact-checking isn't part of the content moderation service, yet.

The Content Moderator service provides answer to these enterprise needs, with the necessary tools for a safe and positive user experience. The service comes with a human review tool and image text and video moderation capabilities, which we will explore in the next sections.

Trying It Out – Building Content Moderators

In this section, we will use the Content Moderator service to show how it can be used in your application. The Content Moderator service has the capabilities to detect questionable content in text, images, and videos, as well as human-in-the-loop capabilities to review these contents. This is a hands-on tour of these capabilities, to enable you to be able to build it as part of your own application. Follow along with these instructions:

1. The easiest way to get started with Content Moderator service is to go to the Content Moderator portal, at `https://contentmoderator.cognitive.microsoft.com/`.

    Figure 6-1 shows the Content Moderator sign-up button.

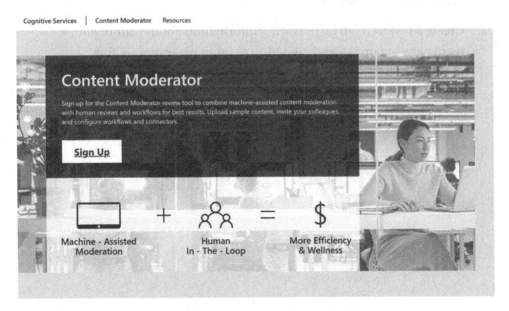

Figure 6-1. *Content Moderator service portal homepage*

2. Click **Sign Up**, and you can log in using your Microsoft account, as shown in Figure 6-2.

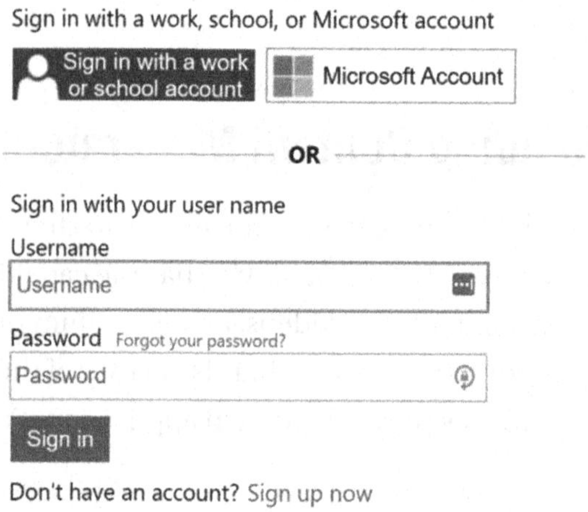

Figure 6-2. *Content Moderator service portal homepage login screen*

This portal is designed for human-in-the-loop review, but you also see the response from the APIs. The next step, after logging in, is to create a review team.

3. Create a review team by providing information about your region, team name, team ID, and invitation to other reviewers, as shown in Figure 6-3. The fields are fairly self-descriptive, and you can always add team members later.

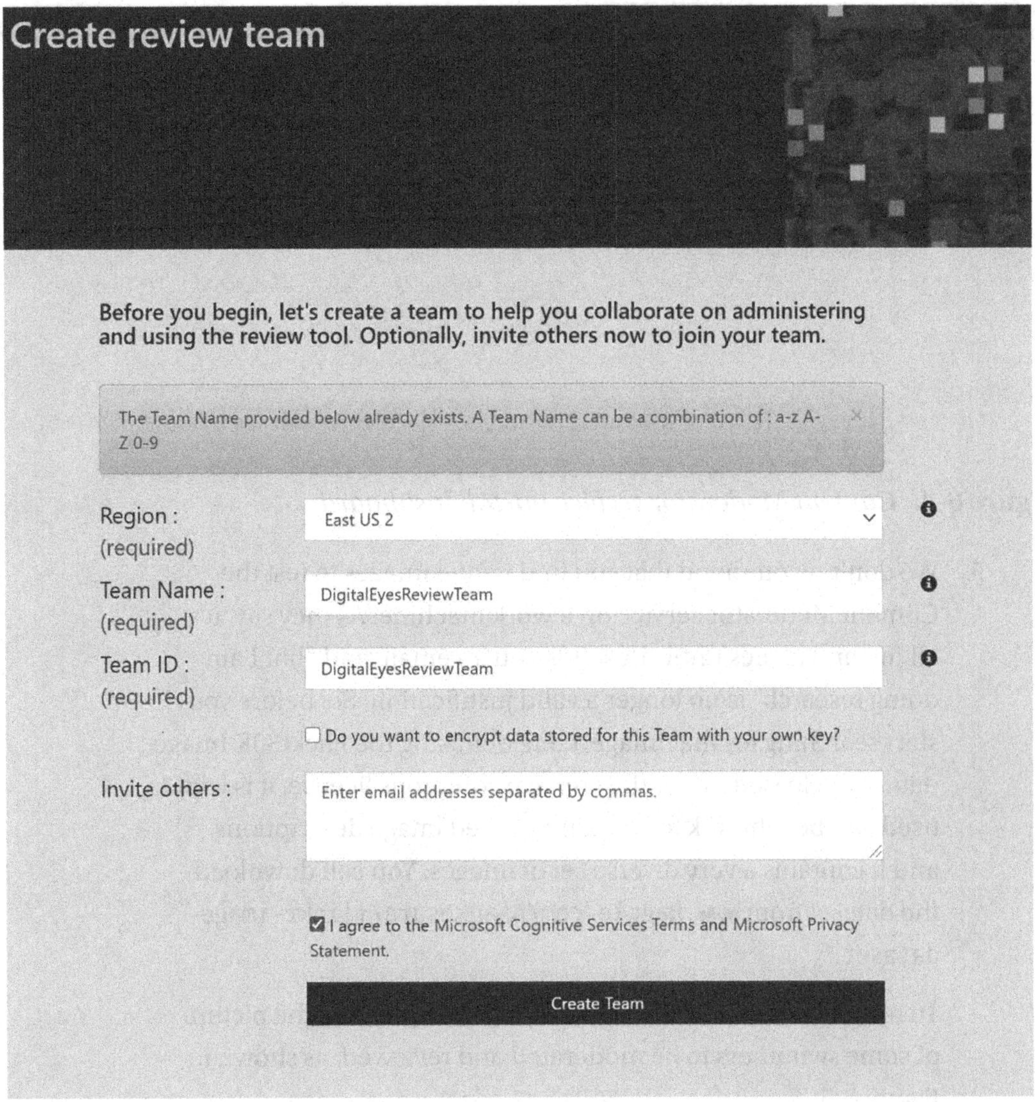

Create review team

Before you begin, let's create a team to help you collaborate on administering and using the review tool. Optionally, invite others now to join your team.

The Team Name provided below already exists. A Team Name can be a combination of : a-z A-Z 0-9   ✕

Region :
(required)

East US 2   ⌄   ❶

Team Name :
(required)

DigitalEyesReviewTeam   ❶

Team ID :
(required)

DigitalEyesReviewTeam   ❶

☐ Do you want to encrypt data stored for this Team with your own key?

Invite others :

Enter email addresses separated by commas.

☑ I agree to the Microsoft Cognitive Services Terms and Microsoft Privacy Statement.

Create Team

Figure 6-3. *Content Moderator service portal Create review team screen*

4. Once you have created the team, you see the dashboard shown in Figure 6-4. This summarized view shows the total requests of submitted review tasks for images, text, and videos, along with all pending and completed requests. You can now try to submit a task, by selecting the right content modality (image, text, or video) from the top menu, as shown in Figure 6-4.

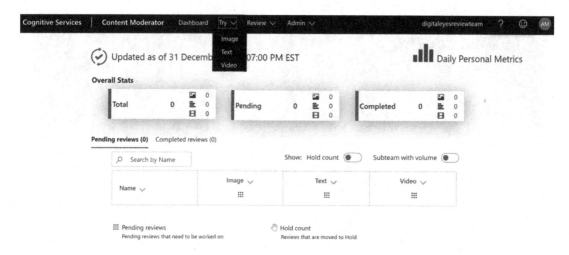

Figure 6-4. *Content Moderator service portal dashboard*

5.   We don't recommend that you find risqué images to test the Content Moderator service on a work machine. As they say, it's all fun and games until HR sends you an email, and "but I am doing research" is no longer a valid justification. So, before you start searching for that image, consider using the Flickr30K Image dataset. Released under the creative commons license, it is widely used as a benchmark for sentence-based image descriptions, and it contains a very diverse set of images. You can download the dataset from `www.kaggle.com/hsankesara/flickr-image-dataset`.

To test the Content Moderator service, we uploaded the picture of some swimmers to be moderated and reviewed, as shown in Figure 6-5.

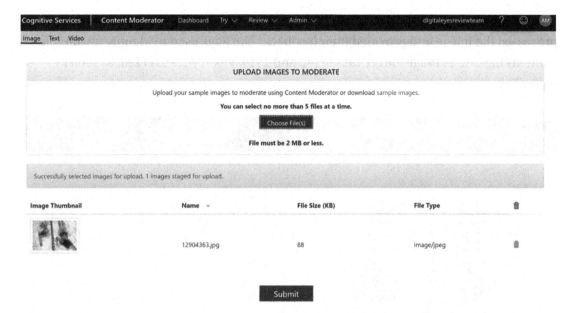

Figure 6-5. *Content Moderator service Upload Images to Moderate screen*

Once you upload the image, you would see the confirmation
shown in Figure 6-6.

Figure 6-6. *Content Moderator service Upload Image Confirmation*

6. Once you get the confirmation, you would see that the image is
pending review, as shown in Figure 6-7.

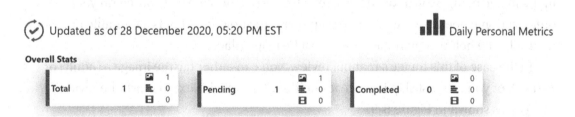

Figure 6-7. *Content Moderator service dashboard status*

As the uploaded image goes through the review workflow, APIs have already processed the image and provided their feedback. The flags used to identify unsuitable contents include isImageAdultClassified and isImageRacyClassified, as seen in Figure 6-8.

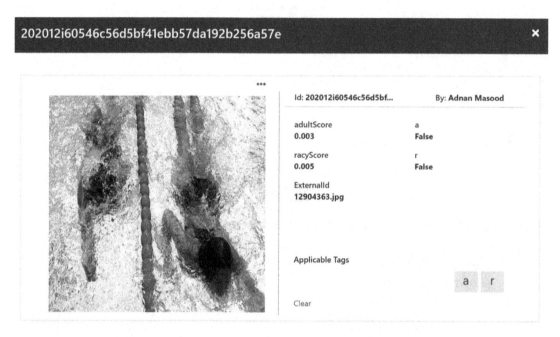

Figure 6-8. Content Moderator service image review scoring screen

The flags are self-explanatory; isImageAdultClassified identifies the sexually explicit images, while isImageRacyClassified flags suggestive comments. The probabilistic score is between 0 and 1, where a higher score means increased confidence in the image being explicit or suggestive in the respective category. Based on your organizational needs, you can set up your own threshold for what is a tolerable number to be allowed before a manual review must be taken place.

In the case of this image's manual review, you can either tag the image a (adult) or r (racy), or you can just skip it, since it's neither. Then click **Next** to finish the review process, as shown in Figure 6-9.

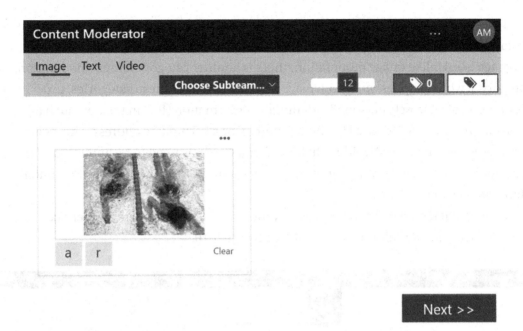

Figure 6-9. *Content Moderator service image review screen*

Now that we have reviewed the human-in-the-loop review scenario of the Content Moderator service, let's explore the text moderation capabilities in the next section.

Moderating Text Using Content Moderator Service

Unstructured data (especially text and natural language) is the largest modality of data available on the Internet. Therefore, the technologies to ingest, consume, process, filter, and prepare to make it insightful are highly sought after. Moderation of text contents, whether they are comments, reviews, or posts in your corporate social media domain, is exactly the kind of problems where the Content Moderator service shines.

The text moderation capabilities of the Content Moderator service are not limited to profanity, but it also helps you classify the questionable language in three different categories. Like the image moderator, the confidence of contents falling in a specific category is between 0 and 1, where higher number demonstrates greater confidence (less of an issue). The ReviewRecommended flag is set to true, in case the service recommends a manual review. You can use this flag, along with the threshold for the category score, to determine whether a review is warranted in your specific case.

The output categories signify different degrees of profane text. Category 1 contains profane, explicit, or adult language, while category 2 implies sexually suggestive language use. Unlike image moderation, there is another category for text moderation, category 3, which implies *mildly* offensive language. Besides these categories, text moderation also detects personally identifiable information (PII) in the text, such as email addresses, IP address, a US phone number, or a US mailing address. This helps in scenarios when personally identifiable information may need to be anonymized or redacted. Auto text correction is another highly useful feature provided by the Content Moderator service.

To get started with this service, we can repeat the content moderation service process as before, but this time with text. See Figure 6-10.

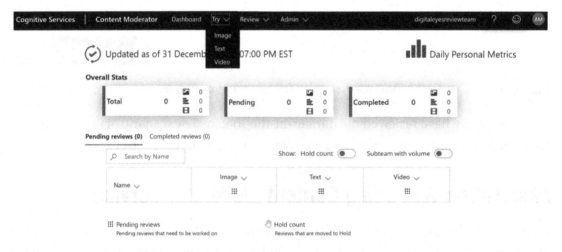

Figure 6-10. *Content Moderator service review screen*

Kaggle, the data science challenge website, provides a Toxic Comment Classification Challenge to identify and classify toxic online comments. The dataset can be downloaded from www.kaggle.com/c/jigsaw-toxic-comment-classification-challenge/data, but we highly recommend not to read through these comments, and they can be very toxic (meaning profane and explicit). We have selected a comment from this dataset to be tested against the service, and you can see it being tried out in Figure 6-11.

Figure 6-11. *Content Moderator service text review screen*

As the confidence numbers show, the comment is neither profane (category 1) nor sexually suggestive (category 2), but the language is mildly offensive or defensive (category 3). The `text.HasProfanity` flag is set to `False`, and so is the `text.hasPII` flag. In the next example, we take a sample text that's provided by the Content Moderator service, and we submit it to be moderated, as shown in Figure 6-12.

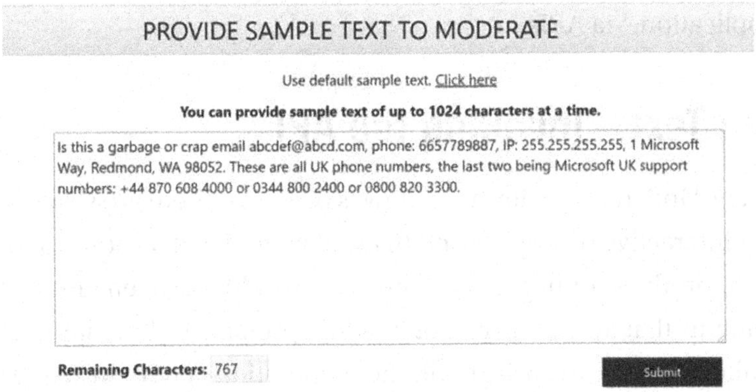

PROVIDE SAMPLE TEXT TO MODERATE

Use default sample text. Click here

You can provide sample text of up to 1024 characters at a time.

Is this a garbage or crap email abcdef@abcd.com, phone: 6657789887, IP: 255.255.255.255, 1 Microsoft Way, Redmond, WA 98052. These are all UK phone numbers, the last two being Microsoft UK support numbers: +44 870 608 4000 or 0344 800 2400 or 0800 820 3300.

Remaining Characters: 767   Submit

Figure 6-12. *Content Moderator service text submission screen*

In this case, the comment has PII information, including an email, phone number, IP address, and US mailing address. This is detected and can be seen in Figure 6-13.

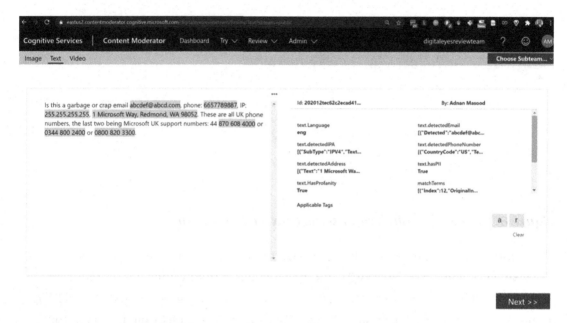

Figure 6-13. *Content Moderator service text submission screen*

In the next section, we will see how to integrate the content moderation capabilities in your own application, via APIs.

Moderating Text – Invoking the API

Using the Content Moderator review workflow is okay, but there are several workflows where you want interactive responses and to build your own decision workflows. For example, for your product support chat rooms, you would like to ensure that any of posted chat contents that are category 1 or 2 (adult, profane, or sexually explicit) are automatically filtered. You can easily build the moderation system by using the Content Moderator APIs.

The following steps show you how to consume the Content Moderator API:

1. First things first, you need to set up a Content Moderator instance on Azure. Visit portal.azure.com and search for "content moderator cognitive service" in the search box, as shown in Figure 6-14.

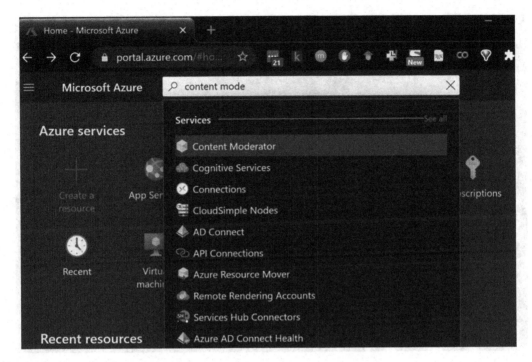

Figure 6-14. *Azure portal Content Moderator service setup*

2. The Content Moderator service displays in the search results, as shown in Figure 6-15. To proceed, click the Content Moderator pane.

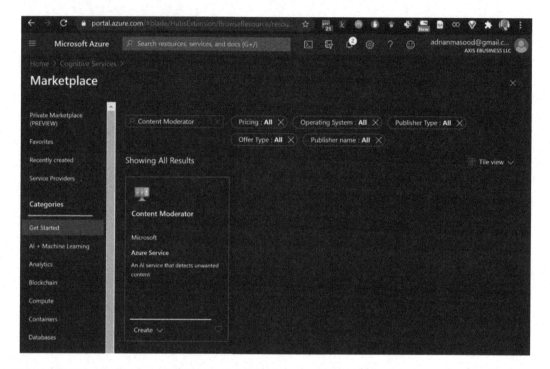

Figure 6-15. *Azure portal Content Moderator service setup*

3. The Content Moderator service detail screen is shown. To proceed, click the **Create** button, as shown in Figure 6-16.

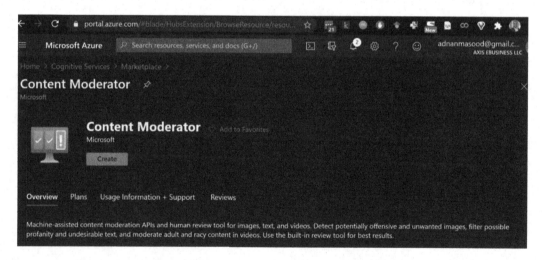

Figure 6-16. *Azure portal Content Moderator service setup*

Now you will create the Content Moderator instance, by selecting the subscription, resource group, instance information, and pricing tier. To continue, click the **Review + create** button, as shown in Figure 6-17.

Figure 6-17. *Azure portal Content Moderator service setup*

The validation occurs, and you see the screen in Figure 6-18 (the validation passed).

Figure 6-18. Azure portal Create Content Moderator screen

Click **Create** to continue deploying the service. You see the screen shown in Figure 6-19, which shows the deployment status.

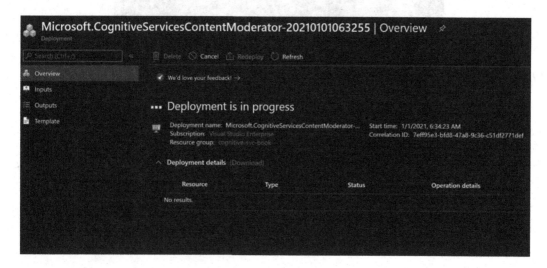

Figure 6-19. Azure portal Deployment is in progress screen

Once the deployment is completed, you are taken to the screen shown in Figure 6-20. To continue, click **Go to resource**.

Figure 6-20. *Azure portal Content Moderator service deployment completed*

The resource screen shows you the Quick start page, where you can get the keys, make API calls, and view the SDK and documentation. See Figure 6-21.

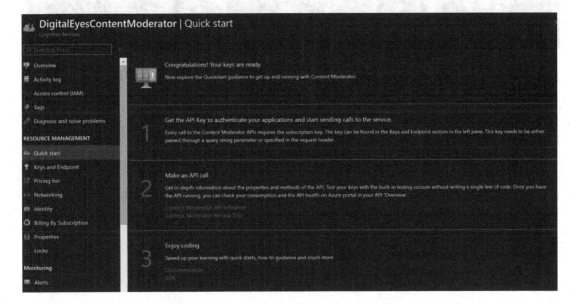

Figure 6-21. *Azure portal Content Moderator Service Quick start*

4. To invoke the API, you need the keys and the endpoint address. Click the **Keys and Endpoint** tab in the left pane. You will see the screen shown in Figure 6-22, with the endpoint and keys.

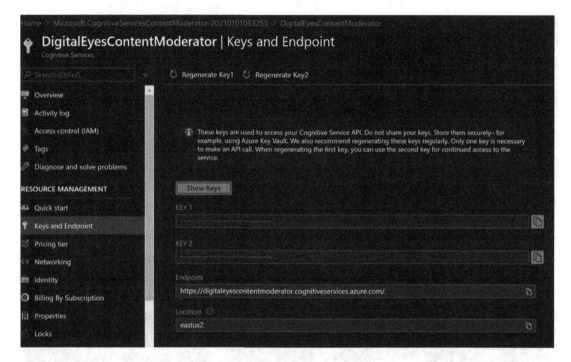

Figure 6-22. *Azure portal Content Moderator service Keys and Endpoint screen*

5. To test the APIs, you can use tools like curl, Postman, Swagger, or
    Fiddler or do it from an application. In this example, we will use
    a .NET application. Cognitive service samples are a great place to
    start exploring different ways of using Azure Cognitive Services.
    First, clone the GitHub repository (`https://github.com/Azure-`
    `Samples/cognitive-services-content-moderator-samples`), as
    shown in Figure 6-23.

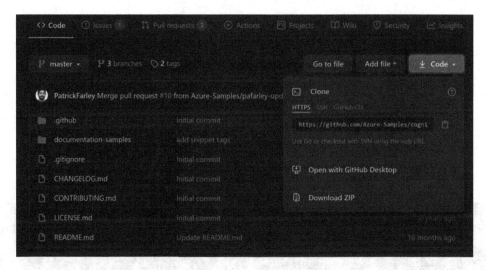

Figure 6-23. *Content Moderator samples GitHub repo*

Clone the repository in a local folder. In this example, we are using GitHub Desktop, but you can do it via command line. The resulting repository is cloned in the following folder (shown in Figure 6-24): `D:\dev\cognitive-services-content-moderator-samples\`.

Clone a repository   ✕

GitHub.com	GitHub Enterprise Server	URL

Repository URL or GitHub username and repository
(`hubot/cool-repo`)

    https://github.com/Azure-Samples/cognitive-services-content-moderator-samples

Local path

    D:\dev\cognitive-services-content-moderator-samples   Choose...

                                          Clone   Cancel

Figure 6-24. *Content Moderator samples GitHub repo clone*

6. Once the repo is cloned (sync'd to your machine), open the
    `text-moderation-quickstart-dotnet.cs` file from the `D:\`
    `dev\cognitive-services-content-moderator-samples\`
    `documentation-samples\csharp` folder. You can open this file
    in Visual Studio Code or in Visual Studio 2019. In this example,
    we will be using Visual Studio and will create a .NET Core 3.1
    console app. We add the code provided in the `text-moderation-`
    `quickstart-dotnet.cs` file, to use it as shown in Figure 6-25.

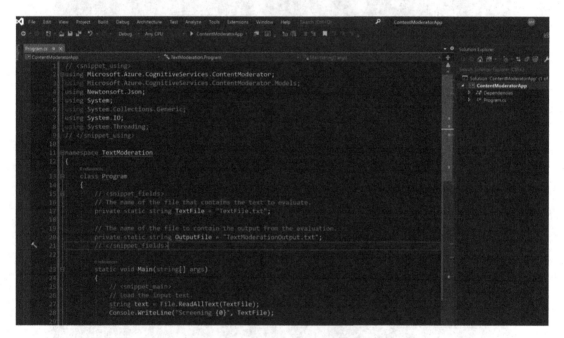

Figure 6-25. *Content Moderator service invocation application*

> *Since the time of this writing, there have been* API *updates, for*
> *instance, Microsoft.CognitiveServices.ContentModerator needs to be*
> *replaced with Microsoft.Azure.CognitiveServices.ContentModerator.*
> *You would get these changes with the latest repo updates.*

7. We need to do a few things before we run this app. First, you
    need to install the Content Moderator packages, using the NuGet
    Package Manager. Run the command shown in Figure 6-26, to get
    the `Microsoft.Azure.CognitiveServices.ContentModerator`
    package.

```
Package Manager   .NET CLI   PackageReference   Paket CLI

PM> Install-Package Microsoft.Azure.CognitiveServices.ContentModerator -Version 2.0.0
```

Figure 6-26. *Command for installation of the NuGet package*

Once the installation is completed, we proceed with configuring the items we need to run the `text-moderation-quickstart-dotnet.cs` file.

8. To run this simple piece of code that calls the Content Moderator API, you need to store the keys in environment variables. This is where this information is read from, as shown in Figure 6-27.

```
// Add your Azure Content Moderator endpoint to your environment variables.
private static readonly string AzureBaseURL =
    Environment.GetEnvironmentVariable("CONTENT_MODERATOR_ENDPOINT");

// Your Content Moderator subscription key.
// Add your Azure Content Moderator subscription key to your environment variables.
private static readonly string CMSubscriptionKey =
    Environment.GetEnvironmentVariable("CONTENT_MODERATOR_SUBSCRIPTION_KEY");
```

Figure 6-27. *Content Moderator invocation – subscription key information*

To add items to the environment variables, open the system properties panel by typing "system properties" on the Run window in Windows 10, as seen in Figure 6-28. Click the **Environment Variables…** button.

Figure 6-28. *System properties to set the environment variables*

Once the window is open, add the two keys (shown in Figure 6-29) as new environment variables. Use user variables instead of system variables. You would need to restart the VS.NET to load the newly added environment variables.

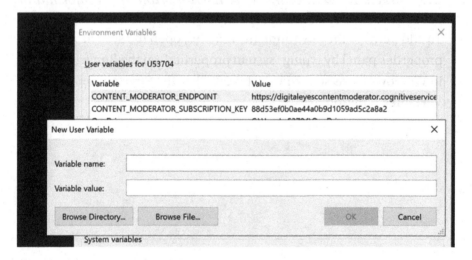

Figure 6-29. *System properties to set the environment variables*

9. Lastly, you would need to create a file that contains the evaluation text to be passed to the API. The text is what we have used earlier. The following is the evaluation text:

```
Is this a grabage or crap email abcdef@abcd.com, phone:
4255550111, IP: 255.255.255.255, 1234 Main Boulevard,
Panapolis WA 96555.
```

Now create a new file called Evaluate.txt, and then copy this text in the file. Make sure to set the copy to output the directory setting to **Copy if newer**, as shown in Figure 6-30. If you don't set this value, the evaluate file is not copied over to the build, and therefore, the executable cannot find the program when it runs.

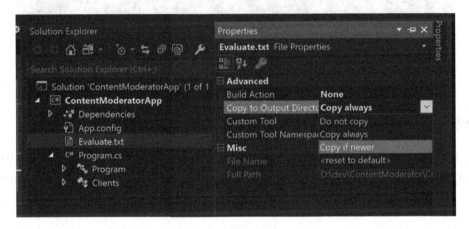

Figure 6-30. *Visual Studio – Copy to Output Directory setting*

10.   Now you can run the program by pressing F5. The code will call the API, get the response back, and save it in the output file, TextModerationOutput.txt, in the bin/debug folder (as shown in Figure 6-31). To enable this view, click the **Show all Files** icon in the Solution Explorer window.

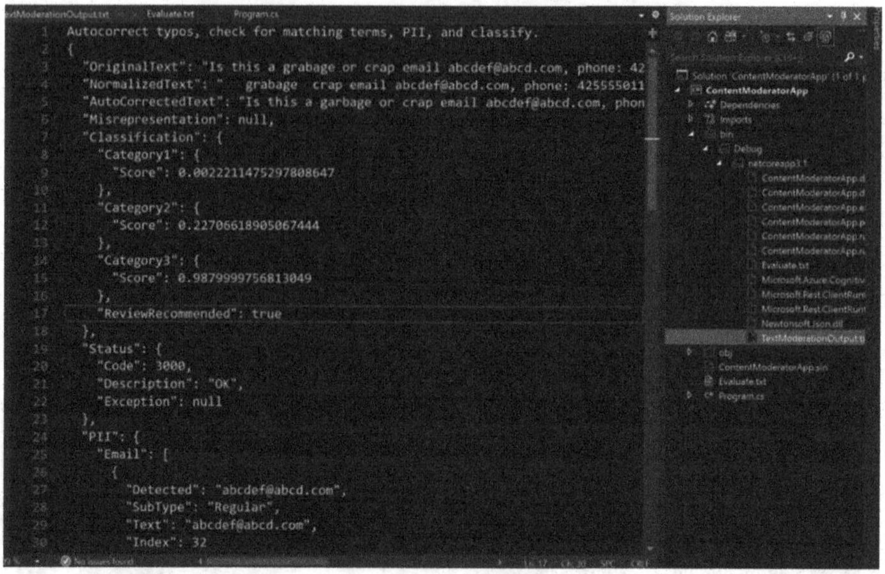

Figure 6-31. *Output listing for the Content Moderator service response*

The JSON output from the API is detailed as follows, where you can easily see the category identification, review recommendation, PII, and language detection – similar to how it was in the review console.

Listing 6-1. Output listing for the Content Moderator service response

```
Autocorrect typos, check for matching terms, PII, and classify.
{
  "OriginalText": "Is this a grabage or crap email abcdef@abcd.com, phone:
  4255550111, IP: 255.255.255.255, 1234 Main Boulevard, Panapolis WA 96555.",
  "NormalizedText": "   grabage  crap email abcdef@abcd.com, phone:
  4255550111, IP: 255.255.255.255, 1234 Main Boulevard, Panapolis WA 96555.",
  "AutoCorrectedText": "Is this a garbage or crap email abcdef@abcd.com,
  phone: 4255550111, IP: 255.255.255.255, 1234 Main Boulevard, Pentapolis
  WA 96555.",
  "Misrepresentation": null,
  "Classification": {
    "Category1": {
      "Score": 0.0022211475297808647
    },
```

```
    "Category2": {
      "Score": 0.22706618905067444
    },
    "Category3": {
      "Score": 0.9879999756813049
    },
    "ReviewRecommended": true
  },
  "Status": {
    "Code": 3000,
    "Description": "OK",
    "Exception": null
  },
  "PII": {
    "Email": [
      {
        "Detected": "abcdef@abcd.com",
        "SubType": "Regular",
        "Text": "abcdef@abcd.com",
        "Index": 32
      }
    ],
    "SSN": [],
    "IPA": [
      {
        "SubType": "IPV4",
        "Text": "255.255.255.255",
        "Index": 72
      }
    ],
    "Phone": [
      {
        "CountryCode": "US",
        "Text": "4255550111",
```

```
        "Index": 56
      }
    ],
    "Address": [
      {
        "Text": "1234 Main Boulevard, Panapolis WA 96555",
        "Index": 89
      }
    ]
  },
  "Language": "eng",
  "Terms": [
    {
      "Index": 12,
      "OriginalIndex": 21,
      "ListId": 0,
      "Term": "crap"
    }
  ],
  "TrackingId": "f0dba865-3731-4fd2-b07f-01878ba7325a"
}
```

Content moderation use cases in the industry exist to provide positive user experiences. They apply to pretty much all the user-generated contents, from moderating text, chat messages, and user-selected usernames (or avatars) to uploaded images that might contain sexually explicit, hateful, violent, or extreme contents, and more. In this section, you have learned how to use the content moderation service with its powerful review features – both via an API and the human-in-the-loop console.

In the next section, we will explore another one of decision services, Personalizer.

Personalizer Service

Even for millennials, it is hard to recognize a digital world without constant recommendations. Blockbuster stores had a hard time keeping the movies clustered in the same genre, while Netflix seamlessly curated to my movie sensibilities. YouTube offers relevant contents and helps me explore similar channels and new artists, while

Spotify constantly digs for all my nostalgic tunes, based on my playlist. Amazon recommends the most relevant products, based on my recent purchases, and financial institutions create customizable offerings, based on my savings and spending patterns. This is hyper-personalization, customization, segmentation, and content curation, designed to create a segment of one (you). It's made possible, due to machine learning approaches that are applied to the problem of serving custom experiences.

To enable and accelerate AI democratization, like all other Cognitive Services, the Personalizer service creates an abstraction layer that allows you, the developer, to create a seamless experience for your customer, without getting into the nitty-gritty of an underlying recommendation engine.

However, if you are interested in the underlying algorithms that are associated with the recommender systems, Microsoft has an excellent GitHub repository (`https://github.com/Microsoft/Recommenders`) with detailed implementations, notebooks, and comparisons of different algorithms. This discussion is beyond the scope of this chapter, but you can find everything you need there, including all the relevant details, associated research papers, Jupyter notebooks, example code, and so on.

The Azure Cognitive Services Personalizer service won the Strata Data Awards for the most innovative product. It can be applied to customizing anything, including but not limited to news articles, products, food, movies, blog posts, and so on. Based on reinforcement learning approaches and customized Microsoft research algorithms, the principle behind the Personalizer service is quite easy to understand. It works like this:

1. Each item (product, news article, food item, movie, and so on) includes attributes or features. The Personalizer API is called and passed these features to the Rank API.

2. Personalizer's underlying engine uses a reward action identifier to determine the best model and recommendation. You determine (train) via user preferences, or via a business rule.

3. Based on this feedback, the reward (re)trains itself by correlating the rank and reward. Now the inference engine has the new model.

Obviously, this is an oversimplification of the process, but if you still feel kind of lost, think of the Personalizer approach as it would operate on the reward behavior. Later, we will cover the details of learning policy, Apprentice mode, exploration vs. exploitation, the trade-off, and so on. But first, it makes sense to do a quick, hands-on experiment and show you how Personalizer works.

Before we start, let us point out that there is a large repository of Personalizer related demos, snippets, and utilities available at `https://github.com/Azure-Samples/cognitive-services-personalizer-samples`. You can also see an interactive working demo at `https://personalizationdemo.azurewebsites.net`, which helps explain the concepts in a visual manner, as shown in Figure 6-32.

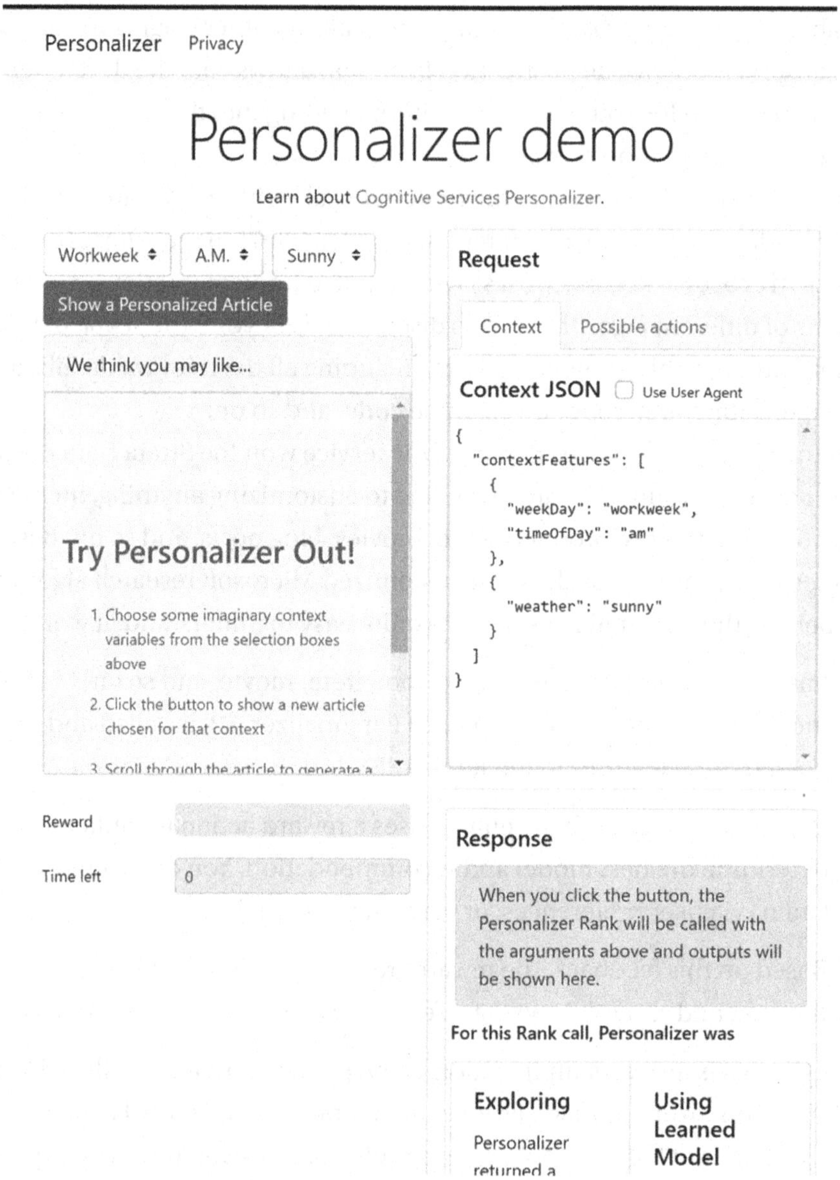

Figure 6-32. *Personalizer services demo*

Personalizer uses Vowpal Wabbit as the foundation for its machine learning implementations. You can read more about it on GitHub at `https://github.com/VowpalWabbit/vowpal_wabbit/wiki`.

Trying It Out – Building a Movie Personalizer

In this example, we will repeat some of the earlier steps of setting up Azure Cognitive Services. Consistent workflow is one of the salient features offered by Cognitive Services, where you can easily replicate your earlier knowledge.

1. Go to `https://portal.azure.com`. Log in with your existing Azure subscription, and then search for Personalizer Service, as shown in Figure 6-33. Click the Personalizer pane.

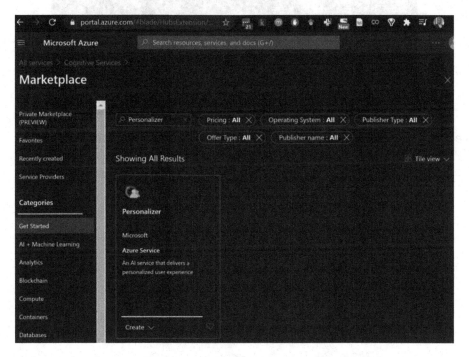

Figure 6-33. *Personalizer service – creating a Personalizer instance*

2. Upon clicking Personalizer, you will see the screen shown in Figure 6-34. Fill in the details to create a Personalizer instance.

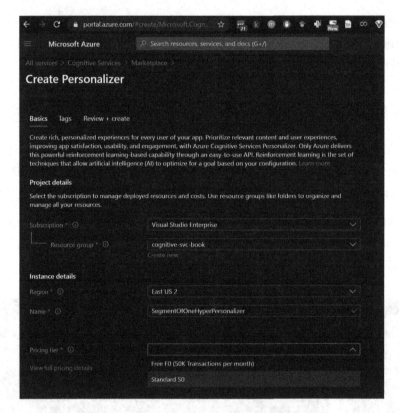

Figure 6-34. *Personalizer service – configuring the Personalizer elements*

The details will be validated, and you would see the screen shown in Figure 6-35. Click **Create** to proceed with your deployment.

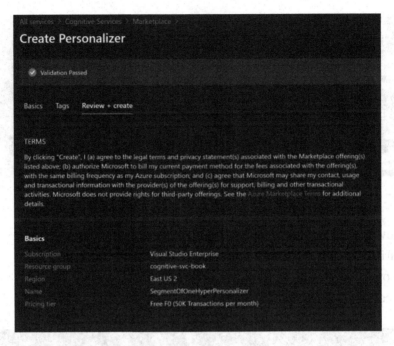

Figure 6-35. *Personalizer service – validation screen*

Once the deployment is in progress, it will deploy the instance, based on the given parameters (see Figure 6-36).

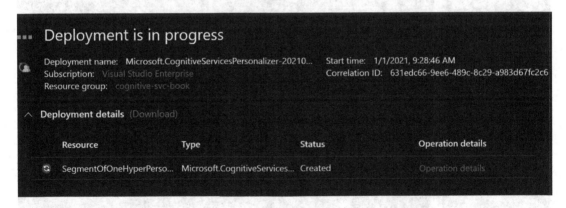

Figure 6-36. *Personalizer service – deployment in progress*

Once the deployment is completed, you will see the screen shown in Figure 6-37. Click the **Go to resource** button to navigate to the next screen.

Figure 6-37. Personalizer service – deployment completed

3.   Now that we have the service deployed, we can consume it via
      various methods that are specified in the Quick start page, as
      shown in Figure 6-38. In our case, we will create a small movie
      recommendation personalizer.

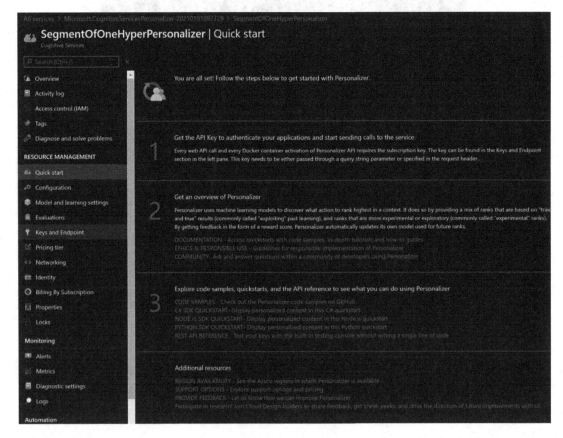

Figure 6-38. Personalizer service – Quick start page

Like the other Cognitive Services, to invoke the Personalizer, you will need the keys and the endpoint. You can find this information by clicking the **Keys and Endpoint** link on the left pane (see Figure 6-39).

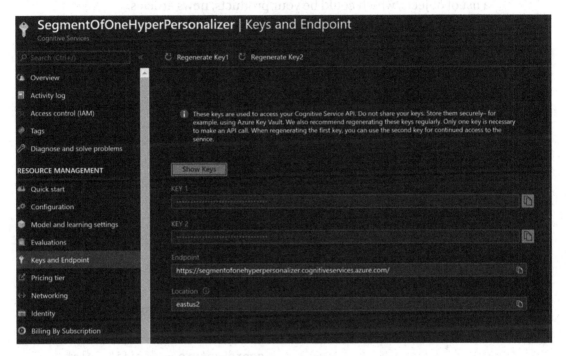

Figure 6-39. *Personalizer service – key and endpoint information*

4. To help explain this concept, let's create a simple application that gets trained. It then recommends a movie, based on genre and popularity. We will define these popularity features as follows:

```
string[] PopularityFeature = {"blockbuster", "indie",
"cult-classic"};
```

We define the genre features as follows:

```
string[] GenreFeatures =
            {"action", "animation", "comedy", "drama",
            "horror", "romance", "scifi", "thriller"};
```

We now need to create a list of `RankableAction` for the movies in our catalogue. Think of this as pretraining for the attributes and features that you have in your dataset. The rankable action takes a list of objects, which could be your products, news articles, or shopping cart items. It then applies the ranking and reward functions, based on the model. In this case, our `RankableAction` list looks like the following, which we have shortened for brevity:

```
IList<RankableAction> actions = new List<RankableAction>
        {
            new RankableAction
            {
                Id = "Interstellar",
                Features =
                    new List<object>
                        {
                            new
                            {
                                imdb = 8.6, rottentomatoes = 72,
                                popularityFeature = "blockbuster",
                                genreFeature = "scifi", plot =
                                "time paradox"
                            }
                        }
            },
            new RankableAction
            {
                Id = "Gattaca",
                Features = new List<object>
                {
                    new
                    {
                        imdb = 7.8, rottentomatoes = 82,
                        popularityFeature = "cult-classic",
                        genreFeature = "scifi",
```

```
                        plot = "genetic profiling"
                    }
                }
            },
        };
```

This is part of the `IList<RankableAction> GetActions()` method. The other two methods are `GetPopularityFeatures()` and `GetGenrePreferences()`, as shown in Figures 6-40 and 6-41.

```
private static string GetPopularityFeatures()

{
    string[] PopularityFeature = {"blockbuster", "indie", "cult-classic"};

    Console.WriteLine(
        "\nWhat popularity feature is it (enter number)? 1. blockbuster 2. indie 3. cult-classic ");
    if (!int.TryParse(GetKey(), out var timeIndex) || timeIndex < 1 || timeIndex > PopularityFeature.Length)
    {
        Console.WriteLine("\nEntered value is invalid. Setting popularity feature value to " +
                          PopularityFeature[0] + ".");
        timeIndex = 1;
    }

    return PopularityFeature[timeIndex - 1];
}
```

***Figure 6-40.** Personalizer service demo – GetPopularityFeatures method*

```
private static string GetGenrePreferences()
{
    string[] GenreFeatures =
        {"action", "animation", "comedy", "drama", "horror", "romance", "scifi", "thriller"};

    Console.WriteLine(
        "\nWhat type of Genre would you prefer (enter number)? 1. action 2. animation 3.comedy 4. drama 5. horror 6. romance 7. scifi 8. thriller");
    if (!int.TryParse(GetKey(), out var tasteIndex) || tasteIndex < 1 || tasteIndex > GenreFeatures.Length)
    {
        Console.WriteLine("\nEntered value is invalid. Setting feature value to " + GenreFeatures[0] + ".");
        tasteIndex = 1;
    }

    return GenreFeatures[tasteIndex - 1];
}
```

***Figure 6-41.** Personalizer service demo – GetGenrePreference method*

5. Now that we have set up the genres, with a set of some movies to train the initial model, we can start invoking the service. But before we do that, let's set up the reward function's wait time to five seconds, which allows for it to be retrained faster. There is a potential performance penalty for retraining so quickly, but for this example, it would be negligible. You can access the following

screen from **Model and learning settings** in the left pane in the main console menu. Set the reward wait time, and then set the model update frequency to five seconds. See Figure 6-42.

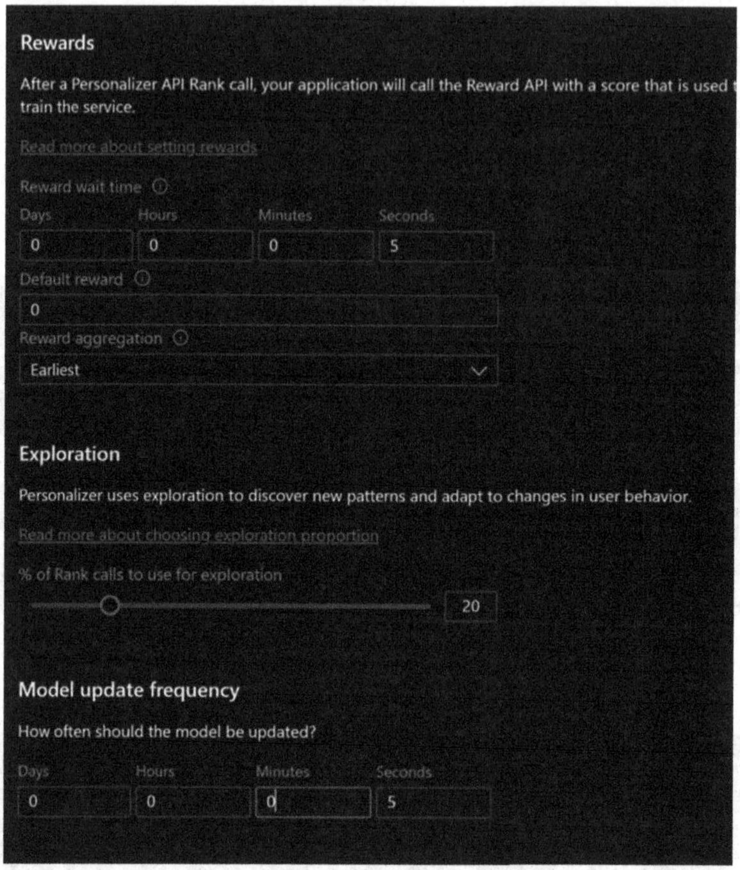

Figure 6-42. *Personalizer service – setting the model update and reward frequency*

6.   To start the ranking and reward process, you can run the program by pressing F5. The entire code list is available as part of **program.cs** file in the Personalizer repository of this book. We will do a walk through of important parts of the code.

First, get the context information, such as the popularity features and movie genre preferences from the users, via the following lines:

```
var popularityFeatures = GetPopularityFeatures();
var genreFeatures = GetGenrePreferences();
```

Next, create a current context, based on the user's data as a list of objects. See the following code:

```
IList<object> currentContext = new List<object>
{
    new {popularity = popularityFeatures},
    new {genre = genreFeatures}
};
```

You can also create a list of excluded actions, any attributes that you want to be considered for ranking. In this case, we don't have any, so we will leave it empty, as follows:

```
IList<string> excludeActions = new List<string> {""};
```

Now, we generate an event ID to associate with the request, as follows:

```
var eventId = Guid.NewGuid().ToString();
```

We then proceed with ranking the actions. The RankRequest() method takes the action, context, and exclusion actions, and then it makes the service call. This call could be part of any of your existing workflows, such as doomsday scroll, serving movie recommendations, generating offers for your new customers, and so on. See the following code:

```
var request = new RankRequest(actions, currentContext,
excludeActions, eventId);
var response = client.Rank(request);
```

The response you get has a reward action ID. Remember, this
is based on the RankableAction set earlier, so it will be a movie
recommendation in our case. See the following line:

```
Console.WriteLine("\nPersonalizer service thinks you
would like to have: " + response.RewardActionId +".
Is this correct? (y/n)");
```

Here, you make the call. If you like the recommendation, you can
adjust the reward accordingly and the model learns, as shown in
the following:

```
if (answer == "Y")
{
    reward = 1;
    Console.WriteLine("\nGreat! Enjoy your movie.");
}
else if (answer == "N")
{
    reward = 0;
    Console.WriteLine("\nYou didn't like the recommended
    movie choice.");
}
```

You also get all the other actions and corresponding probabilities
as part of the response object. See the following code:

```
Console.WriteLine("\nPersonalizer service ranked the
actions with the probabilities as below:");
foreach (var rankedResponse in response.Ranking)
    Console.WriteLine(rankedResponse.Id + " " +
    rankedResponse.Probability);

// Send the reward for the action based on user response.
client.Reward(response.EventId, new RewardRequest(reward));
// </reward>
```

Next, you send the reward for retraining. The process continues to learn, explore, exploit, and retrain as part of the loop, until you exit. This loop is an analogy to how your ecommerce website would have the recommendations system running.

7. You can now run this program to see the code in action. It asks for the popularity feature and genre, and then it lists the recommended movie. During a cold start with limited data, the probabilities of all the products are equally divided. It changes as the Personalizer engine learns. See Figure 6-43.

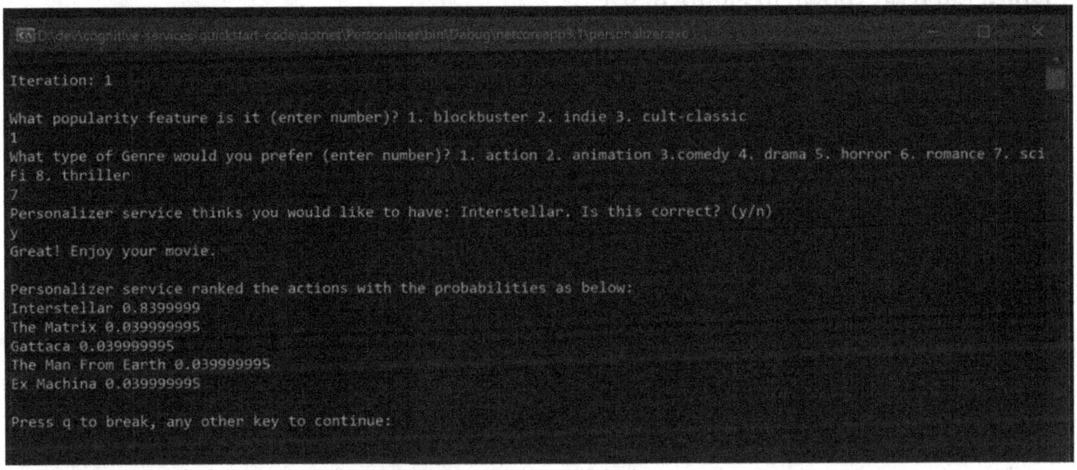

Figure 6-43.   *Personalizer service – ranking and reward demo*

As the results show in Figure 6-43, *Interstellar* is the recommended movie with the highest probability, while rest of the items are shown with their respective rankings. There are multiple ways to train. It includes the Apprentice mode, where you can train the Personalizer without impacting the production application. Learn how to use Apprentice mode to train Personalizer, at `https://docs.microsoft.com/en-us/azure/cognitive-services/personalizer/concept-apprentice-mode`. You can also build feature types to create specific personas for users, based on their attributes. You can then customize the reward functions accordingly.

A very important thing to note is that the resulting recommendation is not always based on the highest ranking. This concept is referred to as the exploration vs. exploitation trade-off. For example, you might love horror movies, but if I just keep recommending *The Exorcist* over and over again, you will definitely miss out on all the other subgenres of horror, which may have ghosts and ghouls, monsters, paranormal activities, and poltergeists. Therefore, it is important to balance between exploration and exploitation in the search space. You should balance between wider-ranging recommendations and exploiting the reward function (where you maximize the reward, based on previous behavior). This approach is used to expand the figurative horizon of the model. You can read more about how to do this with the Anomaly Detector service, at `https://docs.microsoft.com/en-us/azure/cognitive-services/anomaly-detector/overview`.

In the next iteration, I asked our recommender for an indie sci-fi movie, and I got *The Man from Earth*, which is again a good recommendation, even though it's not the highest rated (as shown in Figure 6-44).

```
what popularity feature is it (enter number)? 1. blockbuster 2. indie 3. cult-classic
2
What type of Genre would you prefer (enter number)? 1. action 2. animation 3.comedy 4. drama 5. horror 6. romance 7. sci
Fi 8. thriller
7
Personalizer service thinks you would like to have: The Man From Earth. Is this correct? (y/n)
y
Great! Enjoy your movie.

Personalizer service ranked the actions with the probabilities as below:
The Man From Earth 0.039999995
The Matrix 0.039999995
Gattaca 0.039999995
Interstellar 0.8399999
Ex Machina 0.039999995
```

Figure 6-44. *Personalizer service – ranking and reward demo*

Remember that next time when you see a seemingly unrelated product, it is the algorithm experimenting to broaden its reinforcement learning boundaries. Without that, you would not be exposed to the horror subgenres of zombies, dystopian or apocalyptic worlds, and serial killers.

This example concludes our overview of the Personalizer service. As you can imagine, Personalizer and recommendation systems are everywhere, to capture our attention and to try to understand us better on a personal level. These algorithms are getting really good at it, and as you build your own solution, make sure you consider the ethical ramifications associated with it. Microsoft has provided a sample set of guidelines for responsible implementation of Personalizer at `https://docs.microsoft.com/en-us/azure/cognitive-services/personalizer/ethics-responsible-use`. We highly recommended reading this for everyone building personalization tools and algorithms, either with custom algorithms or via Personalizer and other personalization services.

In the next section, we will review the Anomaly Detector service.

Anomaly Detector Service

Anomalies or outliers are deviations from the usual trends of data. These deviations can be just noise or may hide some very interesting aspects within these unusual data points. This is why finding anomalies has been a topic of significant interest in retail, financial services, healthcare, manufacturing, and cybersecurity verticals. By incorporating historical data, seasonality, and consumer behavioral profiling, effective anomaly detection systems provide greater insights into behaviors that pose a risk to a business or that provide new opportunities for the business.

Simplistic statistical approaches to anomaly detection are typically based on distance-based measures. For example, one approach is to find out how far the data point in question is from the existing distribution, which determines if this is just noise, error in measurement, or something more interesting. The approaches can be roughly classified as predictive-confidence-based approaches, statistical approaches, and clustering-based approaches.

For example, a customer who typically spends only $500 a month on groceries and in supermarkets suddenly doubled his expenditure in March 2020. This outlier gets flagged, but a more sophisticated model will realize that this pattern is emerging globally and is in line with a COVID-19 spending uptick and toilet-paper hoarding. Incorporating seasonality helps make a robust and adjustable model.

For another example, your bank notices a $100 charge on your card for Robux or V-Bucks. Because there is no previous history of this level of expenditure of in-game currency, they know that either somehow your kids have cajoled you into buying them these virtual dollars for their games, or someone is misusing your card. These different types of outliers help identify interesting patterns in data and are very helpful to understand consumer behaviors.

Typical approaches used to determine outliers include z-score (the deviation from the mean value) or extreme value analysis (EVA), linear regression models (such as principal component analysis or least median of square), proximity-based models, information theoretical approaches, or high-dimensional sparse models. Anomaly detection is nuanced, and the Anomaly Detector service abstracts out the details by allowing you to pass the time series data and to identify abnormal behaviors among these data points in an automated manner. The service API provides information about expected value and detects which data points are anomalies (as we will see shortly).

How it works   The explanation is an abstract from the paper

"Time-Series Anomaly Detection Service at Microsoft," by Ren et al. The white paper outlines the algorithms behind a "time-series anomaly detection service that helps customers monitor the time-series continuously and alert for potential incidents on time. In this paper, we introduce the pipeline and algorithm of our anomaly detection service, which is designed to be accurate, efficient, and general. The pipeline consists of three major modules, including data ingestion, experimentation platform, and online compute. To tackle the problem of time-series anomaly detection, we propose a novel algorithm, based on Spectral Residual (SR) and Convolutional Neural Network (CNN). Our work is the first attempt to borrow the SR model from visual saliency detection domain to time-series anomaly detection. Moreover, we combine SR and CNN together to improve the performance of the SR model. Our approach achieves superior experimental results, compared with state-of-the-art baselines on both public datasets and Microsoft production data."

The paper can be downloaded from `https://arxiv.org/abs/1906.03821`.

Microsoft provides an excellent Anomaly Detector demo (`https://algoevaluation.azurewebsites.net/`) to help you understand how this service identifies anomalies from time series data. The API operates on both batch and streaming data, and it identifies the deviations from the normal pattern of events. There are three data samples available. One sample includes seasonality built in, for you to try out different scenarios and discover outliers with different parameters. You can see the demo in action, in Figure 6-45.

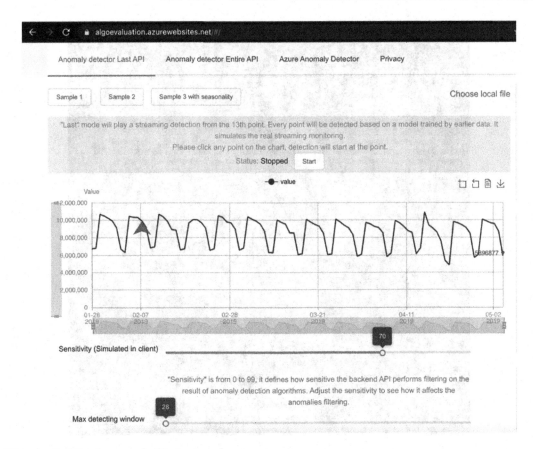

Figure 6-45. *Anomaly Detector demo*

Try It Out – Anomaly Detector Demo

To effectively show the usage of the Anomaly Detector service, we will follow these steps:

> Let's create an Anomaly Detector service instance, as shown in
> Figure 6-46. You would repeat a previous process: starting from
> `https://portal.azure.com`, search for the Anomaly Detector
> service, and then continue creating your work.

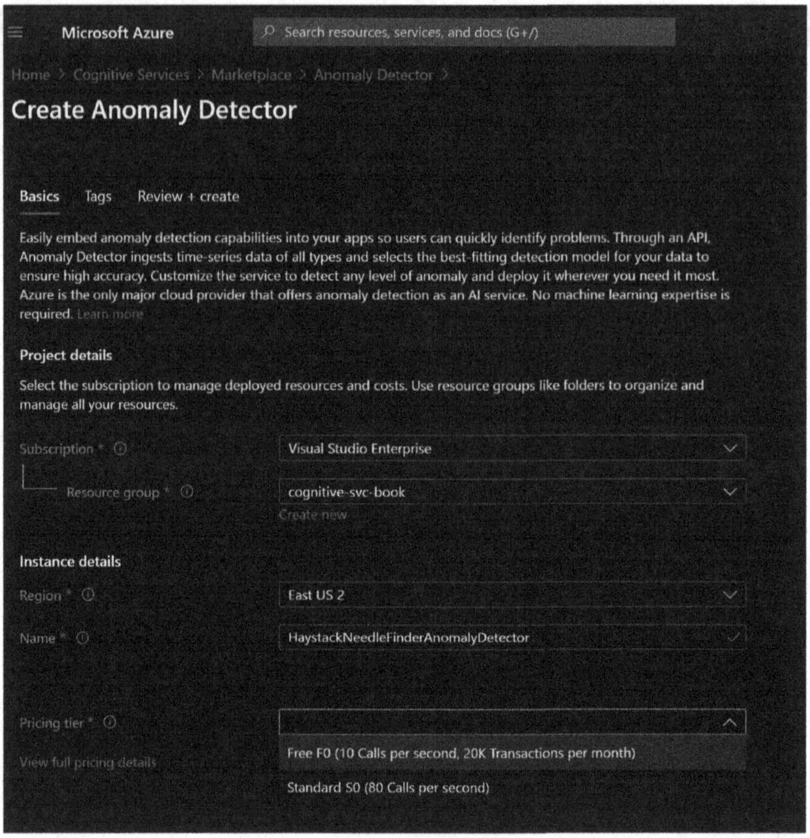

Figure 6-46. *Cognitive Services – setting up Anomaly Detector service*

Set up the Anomaly Detector service parameter, and then click **Create**. The service verifies the parameters, and then create the instance. See Figure 6-47.

Figure 6-47. *Cognitive Services – Anomaly Detector service validation completion*

Once the deployment is completed, you will see the screen in Figure 6-48. Click **Go to resource** to proceed.

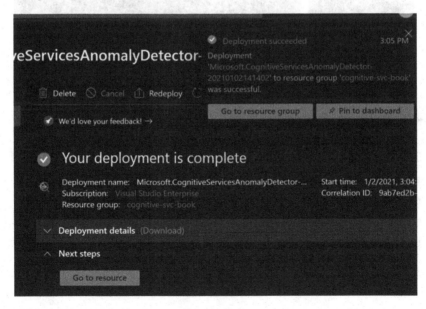

Figure 6-48. *Cognitive Services – Anomaly Detector service deployment completion*

The Quick start page is a great way to get started with the Anomaly Detector service. One of the best ways is to check out the API console shown in section 2 in Figure 6-49. Click the **API Console** link to proceed to the API console.

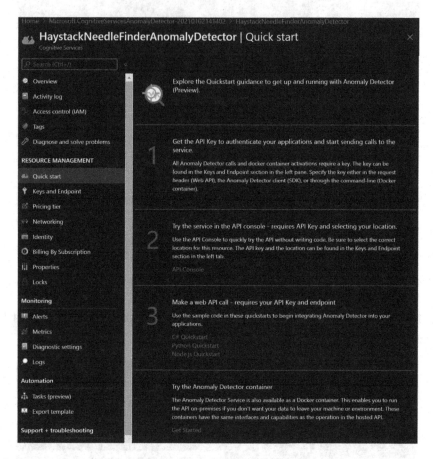

Figure 6-49. *Anomaly Detector service – Quick start*

1. The API console is a web-based UI that demonstrates the capabilities of Anomaly Detector, as shown in Figure 6-50. You can access the console at `https://westus2.dev.cognitive.` `microsoft.com/docs/services/AnomalyDetector/operations/` `post-timeseries-entire-detect/console`.

Figure 6-50. *Anomaly Detector batch invocation*

The console requires the subscription key (as shown in Figure 6-51).
The key can be acquired from the **Keys and Endpoint** link in the
main console.

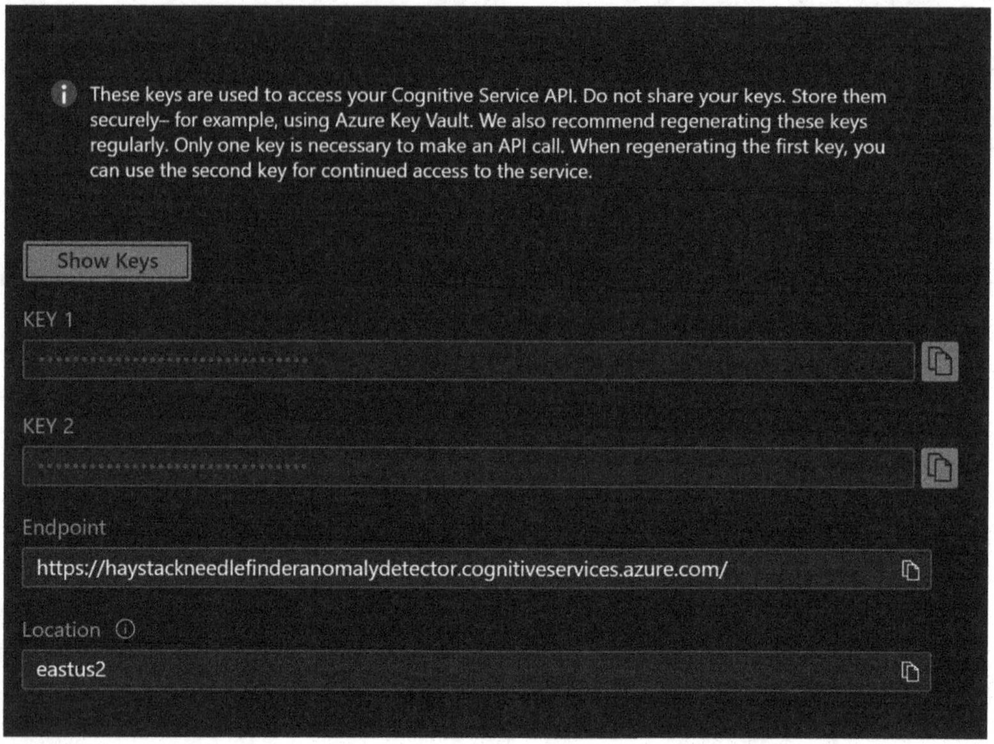

> ⓘ These keys are used to access your Cognitive Service API. Do not share your keys. Store them securely– for example, using Azure Key Vault. We also recommend regenerating these keys regularly. Only one key is necessary to make an API call. When regenerating the first key, you can use the second key for continued access to the service.

Show Keys

KEY 1

KEY 2

Endpoint

https://haystackneedlefinderanomalydetector.cognitiveservices.azure.com/

Location ⓘ

eastus2

Figure 6-51. *Cognitive Services instantiation – keys and endpoint*

The request body is pre-populated with time series data and a set of values. These values are passed over to the Anomaly Detector service as part of the HTTP request (as shown in Figure 6-52).

HTTP request

```
POST https://westus2.api.cognitive.microsoft.com/anomalydetector/v1.0/timeseries/entire/detect HTTP/1.1
Host: westus2.api.cognitive.microsoft.com
Content-Type: application/json
Ocp-Apim-Subscription-Key: ••••••••••••••••••••••••••••••

{
  "series": [
  {
    "timestamp": "1972-01-01T00:00:00Z",
    "value": 826
  },
  {
    "timestamp": "1972-02-01T00:00:00Z",
    "value": 799
  },
  {
    "timestamp": "1972-03-01T00:00:00Z",
    "value": 890
  },
  {
    "timestamp": "1972-04-01T00:00:00Z",
    "value": 900
  },
  {
    "timestamp": "1972-05-01T00:00:00Z",
    "value": 961
  },
  {
```

Figure 6-52. *Anomaly Detector service notebook – request*

2. Once invoked, the Anomaly Detector service does its magic and returns the detailed response, with the expected values and anomalies. The anomalies are directional (such as positive vs. negative anomalies), and you also get to see the margin (ranges) for these outcomes. The details of the Anomaly Detector SDK and API responses can be found at `https://docs.microsoft.com/en-us/dotnet/api/microsoft.azure.cognitiveservices.anomalydetector.models?view=azure-dotnet-preview`. See an example response in Figure 6-53.

Response status

200 OK

Response latency

175 ms

Response content

```
csp-billing-usage: CognitiveServices.AnomalyDetector.DataPoints=1
model-id: 10
x-envoy-upstream-service-time: 33
apim-request-id: 5b29b3b1-6d21-4bc8-89c2-980f9f9c1839
Strict-Transport-Security: max-age=31536000; includeSubDomains; preload
x-content-type-options: nosniff
Date: Sat, 02 Jan 2021 20:51:45 GMT
Content-Length: 3576
Content-Type: application/json

{
  "expectedValues": [827.7940908243968, 798.9133774671927, 888.6058431807189, 900.5606407986661, 962.8389426378304, 9
33.2591606306954, 891.0784104799666, 856.1781601363697, 809.8987227908941, 807.375129007505, 764.3196682448518, 803.9
33498594564, 823.5900620883058, 794.0905641334288, 883.164245249282, 894.8419000690953, 956.8430591101258, 927.628505
5190114, 885.812983784303, 851.7622285698933, 806.3322863536049, 804.8024303608446, 762.74070738882, 804.025170251373
2, 825.3523662579559, 798.0404188724976, 889.3016505577698, 902.4226124345937, 965.867078532635, 937.2113627931791, 8
95.9546789101294, 862.0087368413656, 816.4662342097423, 814.4297745524709, 771.8614479159354, 811.859271346729, 831.8
998279215521, 802.947544797165, 892.5684407435083, 904.5488214533809, 966.8527063844707, 937.3168391003043, 895.18000
3672544, 860.3649596356635, 814.1707285969043, 812.0830500344802, 769.4635044927919, 809.7433654589817],
  "isAnomaly": [false, false, false, false, false, false, false, false, false, false, false, false, false, false, fal
se, false, false, false, false, false, false, false, false, true, false, false, false, false, false, false, false, fa
lse, false, false, false, false, false, false, false, false, false, false, false, false, false, false, false, false],
  "isNegativeAnomaly": [false, false, false, false, false, false, false, false, false, false, false, false, false, fa
lse, false, false, false, false, false, false, false, false, false, false, false, false, false, false, false, false,
false, false, false, false, false, false, false, false, false, false, false, false, false, false, false, false, fals
e, false],
  "isPositiveAnomaly": [false, false, false, false, false, false, false, false, false, false, false, false, false, fa
lse, false, false, false, false, false, false, false, false, false, true, false, false, false, false, false, false, f
alse, false, false, false, false, false, false, false, false, false, false, false, false, false, false, false, false,
false],
  "lowerMargins": [41.389704541219885, 38.91337746719273, 44.43029215903596, 45.02803203993335, 48.14194713189147, 4
6.66295800315348, 44.553920523998386, 42.808908006818456, 40.494936139544734, 40.36875645037526, 4.319668244851755, 4
0.19667492972815, 41.17950310441529, 34.09056413342876, 44.158212262464076, 44.74209500345478, 47.84215295550632, 46.
38142527595062, 44.290649189215177, 42.5881114284947, 40.316614317680205, 40.2401215180422, 2.7407073888200557, 40.201
25851256864, 41.26761831289775, 38.04041887249764, 44.46508252788851, 45.12113062172966, 48.29335392663177, 46.860568
139658994, 44.79773394550648, 43.1004368420068234, 40.82331171048713, 40.72148872762352, 11.8614479159354, 40.59296356
733648, 41.5949913960776, 40.1473772398582, 44.62842203717537, 45.22744107266908, 48.34263531922352, 46.8658419550151
85, 44.759000183627222, 43.018247981783134, 40.70853642984525, 40.604152501724, 9.463504492791913, 40.48716827294913
4],
```

Figure 6-53. *Anomaly Detector service notebook – response*

3. Unlike the previous two examples, where we created the solution
    as a .NET application, we will expose you to the world of Jupyter
    notebooks with Anaconda. Notebooks are to data scientists like
    an IDE to software developers. Many software engineers are
    switching to notebooks for their Python work, due to their ease of
    use and statelessness.

    Microsoft offered its own version of notebooks called Azure
    notebooks (https://notebooks.azure.com/), which is retired. Now
    you will be able to use notebooks with Azure Machine Learning
    (see https://docs.microsoft.com/en-us/azure/notebooks/
    quickstart-export-jupyter-notebook-project#use-notebooks-
    with-azure-machine-learning). For this experiment, we will use

Anaconda-based Jupyter notebooks. Start by downloading and installing Anaconda from `www.anaconda.com/products/individual`. Once installation is complete, you will see a screen similar to the one in Figure 6-54.

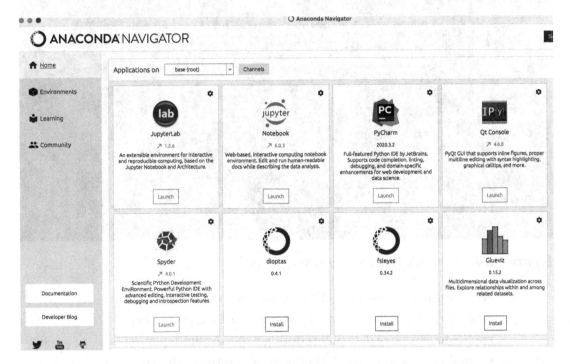

Figure 6-54. *Anaconda Navigator environment*

Continue by clicking the **Launch** button on the JupyterLab pane, which opens the Jupyter environment.

4. For this example, we will use the Anomaly Detector notebook provided with Azure Samples. You can clone the GitHub repository from `https://github.com/Azure-Samples/` `AnomalyDetector`. The repository contains example data, Python notebooks, and Quick start samples. Once cloned, navigate to the folder `\AnomalyDetector\ipython-notebook` and open the file `Batch anomaly detection with the Anomaly Detector API.` `ipynb` from the left pane (as shown in Figure 6-55).

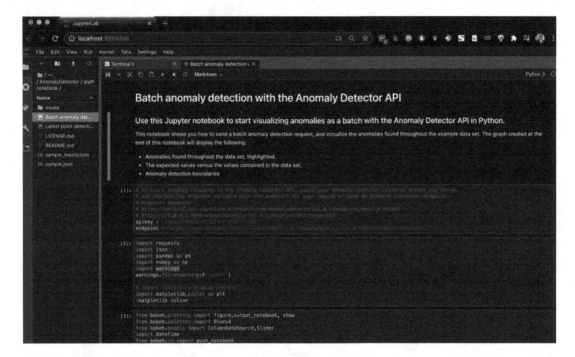

Figure 6-55. *Anomaly Detector service notebook*

5. With the Anomaly Detector API, you can perform batch and point
    (streaming) identification in time series data. Batch detection
    applies to a batch of data points during a specific time span, while
    streaming outliers are continuously monitored on each data
    point. The Anomaly Detector API is a stateless service, and its
    performance is heavily driven by the time series data preparation,
    the API parameters used, and the number of data points. In the
    case of this example, we set up all the prerequisites, including
    Bokeh, the interactive visualization library, as shown in Figure 6-56.

```
[3]:  from bokeh.plotting import figure,output_notebook, show
      from bokeh.palettes import Blues4
      from bokeh.models import ColumnDataSource,Slider
      import datetime
      from bokeh.io import push_notebook
      from dateutil import parser
      from ipywidgets import interact, widgets, fixed
      output_notebook()

      BokehJS 1.4.0 successfully loaded.
```

Figure 6-56. *Anomaly Detector service notebook – initializing prerequisites*

6. Once the prerequisites are all set, you call the API via the detect()
    method, which takes the dataset, subscription key, and endpoint,
    and then it invokes the service. Figure 6-57 shows the description
    of the method.

```
[4]: def detect(endpoint, apikey, request_data):
         headers = {'Content-Type': 'application/json', 'Ocp-Apim-Subscription-Key': apikey}
         response = requests.post(endpoint, data=json.dumps(request_data), headers=headers)
         if response.status_code == 200:
             return json.loads(response.content.decode('utf-8'))
         else:
             print(response.status_code)
             raise Exception(response.text)
```

Figure 6-57. *Anomaly Detector service notebook – the detect method*

7. Once completed, run the hourly sample frequency analyzer (cell
    #6), as shown in Figure 6-58. The build figure method takes two
    parameters, the sample data and the sensitivity, and then it plots
    the outliers with a boundary, expected value, and actual value.

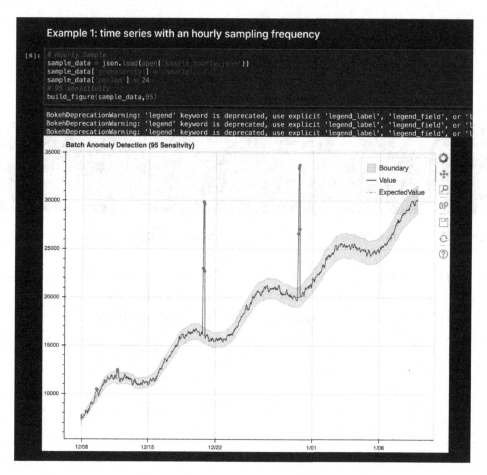

Figure 6-58. *Anomaly Detector service – hourly sampling frequency*

8. Similar to the hourly sample, you can perform time series with
    a daily sampling frequency, by invoking cell #9, as shown in
    Figure 6-59. Here, we set the granularity to be daily (as compared
    to hourly) and build the figure. It shows the daily outliers within
    the dataset, along with the associated boundary, value, and
    expected values.

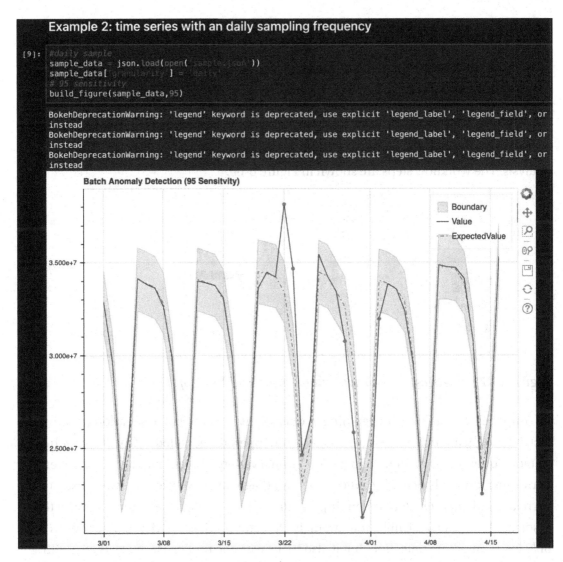

Figure 6-59. *Anomaly Detector service – daily sampling frequency*

This concludes our coverage of the Anomaly Detector service. There is a lot more information about Anomaly Detector that you can read in Microsoft Docs, at `https://docs.microsoft.com/en-us/azure/cognitive-services/anomaly-detector`.

Metrics Advisor (Preview)

One of the newest members of Azure Cognitive Services is Metrics Advisor (see https://
azure.microsoft.com/en-us/services/cognitive-services/metrics-advisor). It is
a special case of outlier analysis services. Metrics Advisor is in preview at the time of this
writing, which means the screens and underlying services may change. Metrics Advisor
is a cognitive service that helps monitor different metrics, and it assists with root cause
analysis. The workflow steps are shown in Figure 6-60.

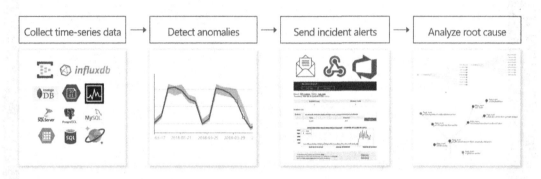

Figure 6-60. *Metrics Advisor workflow, courtesy Microsoft Docs*

Applying AI and machine learning to operations is an area of tremendous interest. In
fact, it brought about a whole new subfield in IT operations, called AIOps. That's where
AI techniques are used for data ingestion, from a variety of event management systems,
performing anomaly detection, and respective diagnostics. Metrics Advisor is used to
correlate and analyze data from multiple sources. It then diagnoses anomalies and the
associated root causes. Unlike the Anomaly Detector service, Metrics Advisor doesn't
only focus on outlier detection, but it also helps with the root cause analysis and with
incident alert management.

1. Like all of Cognitive Services, we start by creating a Metrics
   Advisor instance in portal.azure.com, and we set the parameters.
   See Figure 6-61.

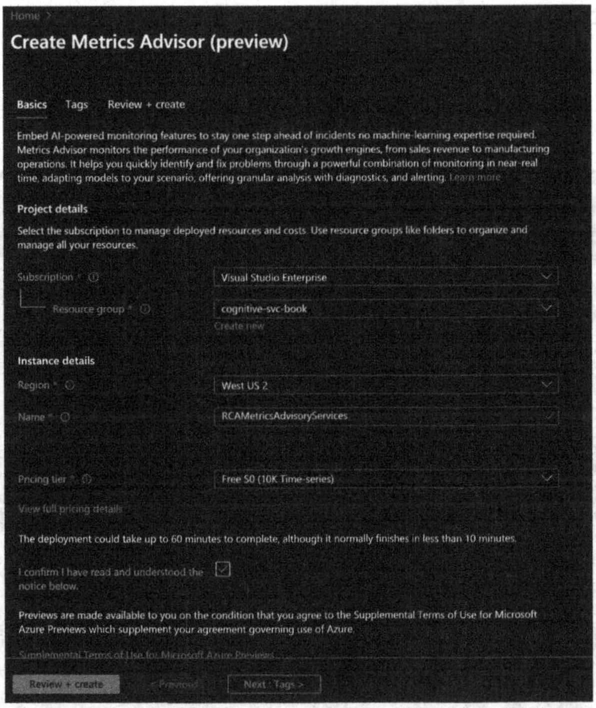

Figure 6-61. *Metrics Advisor console – setup*

In this case, the deployment takes a while (about 22 minutes) to complete, possibly because the service is still in preview phase. See Figure 6-62.

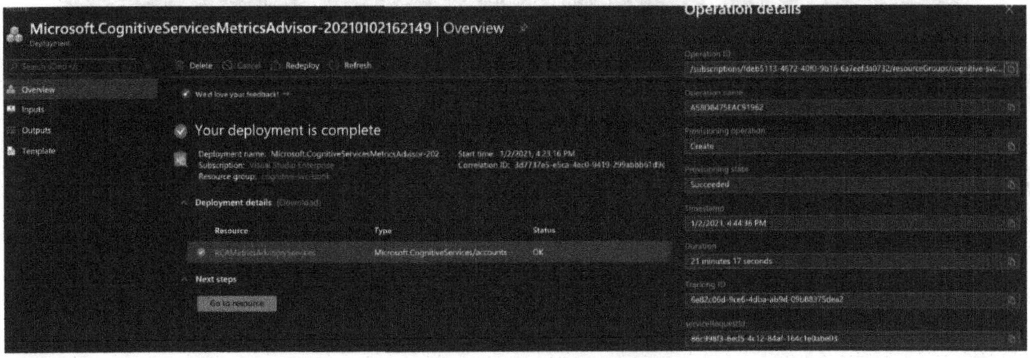

Figure 6-62. *Metrics Advisor console – deployment completed*

Click **Go to resource** to open the Quick start page, which shows you multiple options of what you can do with the Metrics Advisory service. Click **Go to your workplace** in the first step. See Figure 6-63.

Figure 6-63. *Metrics Advisor console – Quick start*

2. This should take you to `https://metricsadvisor.` `azurewebsites.net`, where you can log in and create your own advisor (as shown in Figure 6-64). On this page, you would specify the directory (organization), subscription, and associated workspace.

Figure 6-64. *Metrics Advisor service console*

3. Once completed, you will walk through a detailed Metrics Advisor step-by-step tour, to get you familiar with the environment. This includes the creation of the data source, as shown in Figure 6-65.

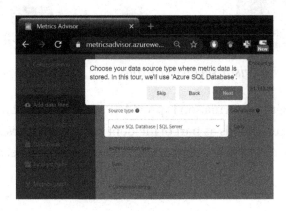

Figure 6-65. *Metrics Advisor service console*

Next, you select the granularity of the metrics (the time period), as shown in Figure 6-66.

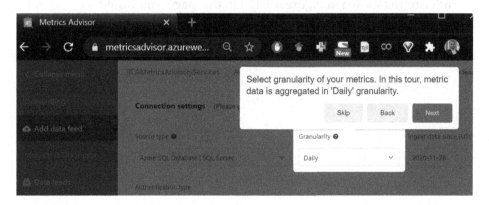

Figure 6-66. *Metrics Advisor service console – setup*

You can perform a basic configuration of the time series data, such as the first available time stamp (see Figure 6-67).

Figure 6-67. *Metrics Advisor service console*

Once this initial setup information is provided, you can now consume and interact with the data feeds, create feeds as part of incident hub, see metrics graphs, create data hooks, and access the API keys to consume this data from a service. See Figure 6-68.

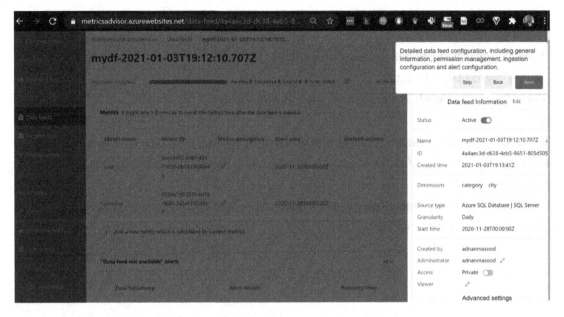

Figure 6-68. *Metrics Advisor service console*

The Metrics Advisor portal helps you onboard the metric data (such as event logs and SIEM data), provide diagnostic insights, visualize the metrics, improve the detection configuration for specific use cases, and create alerts for outliers.

This new but comprehensive Metrics Advisor service is evolving, based on customer needs and feedback. It will soon graduate to become a full member of the amazing Cognitive Services suite. Stay tuned.

Summary and Conclusion

In this chapter, you have developed a basic understanding of how decision-based Cognitive Services and Decision APIs work. You have created a content moderator application, personalized experiences with the Personalizer, reviewed time series anomalies with the Anomaly Detector, and explored the AIOps landscape with the Metrics Advisor service. As you may realize, it is quite difficult to cover these services in any significant level of detail. Therefore, we apologize for skimming over some of the things (to keep it relatively brief). Rest assured, the links and Microsoft documentation have more than ample information for you to explore further.

Now it's your turn – we cannot wait to hear how you are using these services within your organizations, to make important decisions!

References and Further Reading

Azure Cognitive Services documentation: https://docs.microsoft.com/en-us/azure/cognitive-services

Time-Series Anomaly Detection Service at Microsoft: https://arxiv.org/abs/1906.03821

Guidelines for responsible implementation of Personalizer: https://docs.microsoft.com/en-us/azure/cognitive-services/personalizer/ethics-responsible-use

Personalizer documentation: https://docs.microsoft.com/en-us/azure/cognitive-services/personalizer

CHAPTER 7

Search – Add Search Capabilities to Your Application

How old was master Yoda when he died?

How many quarts are in a liter?

How late does the boba place on Fowler stay open?

Okay, Google, make cat sounds.

Well, that last request might not be very relevant. Anyway, it's quite evident from our everyday behavior that digital search has become an extension of the modern human mind. Search is everywhere – in our personal lives, to the companies we build and work for; we are constantly searching for information in our everyday lives. Search has become sophisticated, as the information we seek becomes more diverse, multimodal, and distributed. We now heavily rely on context-sensitive, natural language searches, where answers span across multiple documents (including images, PDFs, text files, and structured data). Our searches also include people and places, individuals and business names, and other entities, which likely require autocorrection for spelling and intent.

By asking questions, we explore the Internet via search engines, we probe for data points, and we seek tutorials, guides, and how-to videos. We also query our corporate intranet portals for holiday/vacation requests and co-payment amounts, we hunt for medical provider coverage and locations, and we investigate which legal clauses are part of a specific contract. Even though we are becoming quite good a probing for all this information, the data itself doesn't automatically lends itself to be examined that easily.

© Ed Price, Adnan Masood, and Gaurav Aroraa 2021
E. Price et al., *Hands-on Azure Cognitive Services*, https://doi.org/10.1007/978-1-4842-7249-7_7

According to most estimates, over 80% of organizational data is unstructured. (See "What Is Unstructured Data and Why Is It So Important to Businesses?" at `www.forbes.com/sites/bernardmarr/2019/10/16/what-is-unstructured-data-and-why-is-it-so-important-to-businesses-an-easy-explanation-for-anyone`.) All this unstructured data leads to practical challenges of information extraction, such as parsing, optical character recognition, named entity recognition, dependency parsing, and other data enrichment needs. But fear no more, Microsoft's AI-powered search is here to help!

In this chapter, we will provide insights on Bing Search APIs and Azure Cognitive Search, by adding various search functionalities to the application. In this chapter, you will learn to do the following:

1. Understand search, Bing Search APIs, and Cognitive Search.

2. Create smart search applications by adding Bing Search Azure Cognitive Search capabilities.

Let's begin.

The Search Ecosystem

Microsoft's search ecosystem has gone through a few iterations that can be easily explained, as follows.

Azure Cognitive Search, formerly known as Azure Search, is Microsoft's enterprise search capability (where you can choose your own data sources to crawl and index). Therefore, even though it could technically encompass web searches, it's more than that. You can also choose how often you index. Essentially, you can create your own mashup of data sources to index from.

Bing Search is specifically Microsoft's enterprise web search. You can create silos of domains from the web that you want to allow your enterprise to search. Azure Cognitive Search is a custom enterprise search. It can be siloed to on prem, private, and enterprise-specific data, with the option to include a custom public domain. In contrast, Bing Search is a public domain that can be filtered.

Currently, Bing Search APIs are listed as part of Azure Cognitive Services, which can create confusion. However, Microsoft has announced that Bing Search APIs will transition from Azure Cognitive Services to Azure Marketplace on October 31, 2023[1], which will help address the separation of concerns. According to the announcement, *"Bing Search APIs provisioned using Cognitive Services will be supported for the next three years or until the end of your Enterprise Agreement, whichever happens first."*

Azure Cognitive Search

Formerly known as Azure Search, Azure Cognitive Search is the AI-powered search service with extensive data enrichment capabilities, including OCR, NER, and key phrase extraction. Provided as a fully managed service, Azure Cognitive Search is supported by decades of work done at Microsoft Research on NLP, Office, Bing, and other search solutions. With crawling capabilities on some data sources, Azure Cognitive Search offers geospatial search, filtering, autocomplete, and search stacking or faceting features. Custom models help support your subject matter, as well as domain-specific client needs. More on the service offering can be read at `https://azure.microsoft.com/en-us/services/search/`.

Searching with Azure Cognitive Search

1. In the Azure portal, search for Azure Cognitive Search, and then click the "Azure Cognitive Search" search result, as it appears in the drop-down menu (shown in Figure 7-1). Select Azure Cognitive Search.

---

[1] `https://azure.microsoft.com/en-us/updates/bing-search-apis-will-transition-from-azure-cognitive-services-to-azure-marketplace-on-31-october-2023/`

Figure 7-1. *Microsoft Azure Portal Offerings search results*

2. The screen in Figure 7-2 shows all the capabilities and features of Azure Cognitive Search. As discussed earlier, this includes scalability, manageability, feature enrichment (OCR, NER, faceting), and custom models. Click Create to proceed.

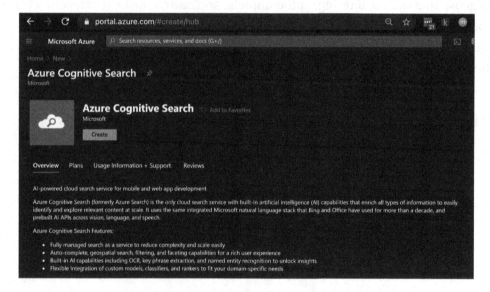

Figure 7-2. *Microsoft Azure Cognitive Search – create offering screen*

3. Fill out the information to create the service. This includes subscription information, the unique URL, resource group, location, and the pricing information (as shown in Figure 7-3).

Figure 7-3. *Azure Cognitive Search – create instance*

The following pricing information (Figure 7-4) is for different tiers. It is subject to change and is provided here as a sample reference only. Please review the current pricing information before use.

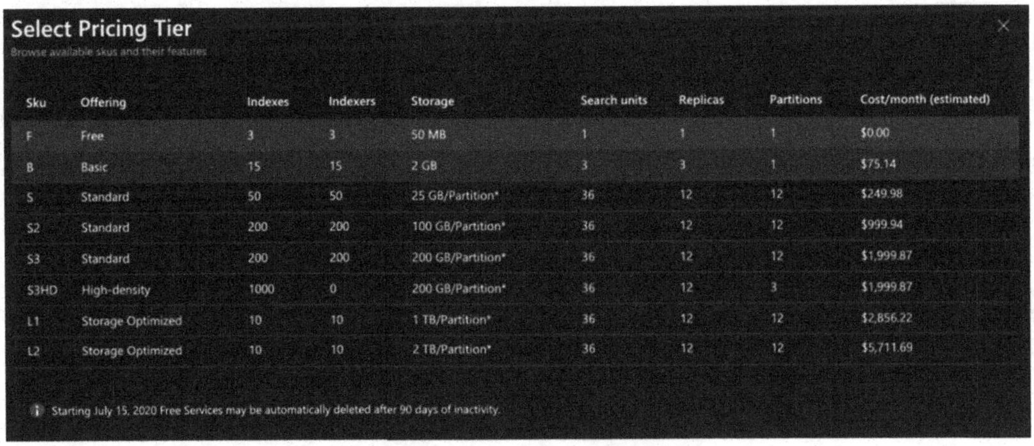

Select Pricing Tier

Browse available skus and their features

Sku	Offering	Indexes	Indexers	Storage	Search units	Replicas	Partitions	Cost/month (estimated)
F	Free	3	3	50 MB	1	1	1	$0.00
B	Basic	15	15	2 GB	3	3	1	$75.14
S	Standard	50	50	25 GB/Partition*	36	12	12	$249.98
S2	Standard	200	200	100 GB/Partition*	36	12	12	$999.94
S3	Standard	200	200	200 GB/Partition*	36	12	12	$1,999.87
S3HD	High-density	1000	0	200 GB/Partition*	36	12	3	$1,999.87
L1	Storage Optimized	10	10	1 TB/Partition*	36	12	12	$2,856.22
L2	Storage Optimized	10	10	2 TB/Partition*	36	12	12	$5,711.69

ⓘ Starting July 15, 2020 Free Services may be automatically deleted after 90 days of inactivity.

Figure 7-4. *Azure Cognitive Search – Select Pricing Tier*

Note   It is important to ensure that your search instance and the Cognitive Services instance used to enrich the data are in the same region. If you'd like your service to be migrated, make sure that the offerings are supported in the target region[2].

4.   You will see screen from Figure 7-5, once you are done creating the service. Click **Go to resource** to see the complete dashboard.

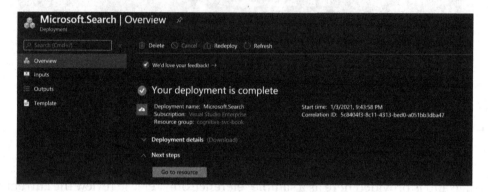

Figure 7-5. *Azure Cognitive Search – deployment completed*

The Cognitive Services dashboard (shown in Figure 7-6) includes the details of quotas, service status, location, subscription information, location, and usage information. This dashboard comes in quite handy, to monitor how storage and indexes perform with additional data sources.

---

[2]Move your Azure Cognitive Search service to another Azure region, `https://docs.microsoft.com/en-us/azure/search/search-howto-move-across-regions`

Figure 7-6. *Azure Cognitive Search – dashboard*

5. Now that we have created a search service instance, let's import
    some data to be searched. Click **Import data** from the top menu
    in the dashboard, and you will see the screen (as shown in
    Figure 7-7) to connect to your data. Complete the following four
    steps:

    a. Connect to the data source (such as Azure SQL Database,
        SQL Server on Azure VMs, Cosmos DB, Azure Blob Storage
        for documents and PDFs, Azure Data Lake Storage Gen2, or
        Azure Table Storage).

    b. Add Cognitive Skills for enrichment (such as NER, key phrase
        extraction, and so on).

    c. Customize the Target Instance.

    d. Finally, create an indexer.

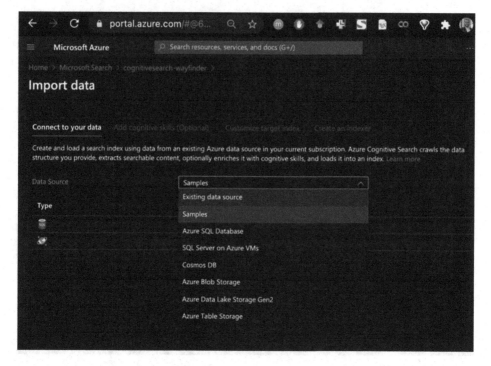

Figure 7-7. *Azure Cognitive Search – Import data*

To keep this simple, we will import data from the samples, specifically the real estate sample (as shown in Figure 7-8).

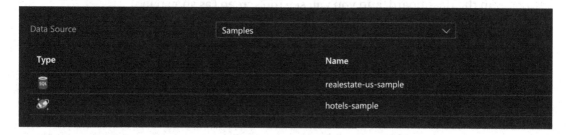

Figure 7-8. *Azure Cognitive Search – import the dataset*

The data comes from SQL and contains real estate information about homes. The next step is to add enrichment or cognitive skills, as shown in Figure 7-9. In this case, we will perform some limited enrichments to recognize the entities.

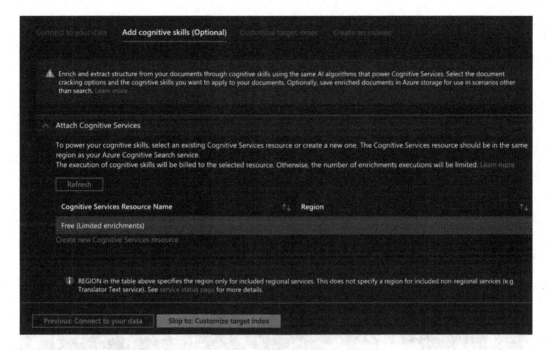

Figure 7-9. *Azure Cognitive Search – Add cognitive skills*

The Cognitive Services enrichments include extracting people, organization, and location names, detecting key phrases, and extracting personally identifiable information (PII). You can create custom enrichments and add domain-specific skills to recognize your industry keywords and phrases. For instance, the ICD-10-CM/PCS Medical Coding Reference can be mapped to corresponding diagnostic and procedure keywords. There is a lot you can do to build Custom Skills in an Azure Cognitive Search enrichment pipeline. You can read the details about attaching a Cognitive Services resource to a skillset in Azure Cognitive Search at `https://docs.microsoft.com/en-us/azure/search/cognitive-search-attach-cognitive-services#limits-when-no-cognitive-services-resource-is-selected`. For this exercise, we will perform the named entity recognition enrichments only. See Figure 7-10.

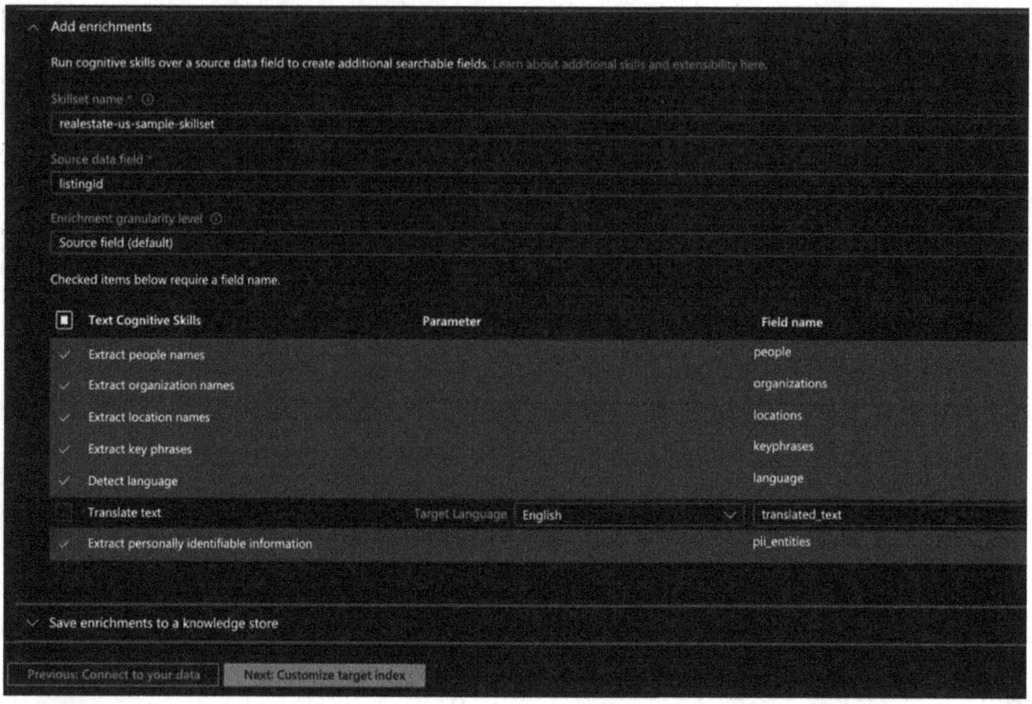

Figure 7-10. *Azure Cognitive Search – adding cognitive skill enrichments*

After adding cognitive skill enrichments, the next step in your Azure Cognitive Search setup is to create a target index. If you are familiar with other search programs, such as Apache Solr, Apache Lucene, or Elasticsearch, you might know that indexing helps you speed up the search. It stores the contents used for full text search, so that it's optimized for faster searches with the physical schema. In Figure 7-11, you'll find the recommended index, along with fields listed that are retrievable, filterable, sortable, facet-able, and searchable. This target index gives you excellent control over your data, so you can granularly include or exclude the attributes to be included in the search. For numerical values, like the number of bedrooms and bathrooms, you can request for it to be sortable. Microsoft has made it very easy with default values preselected, as shown in Figure 7-11.

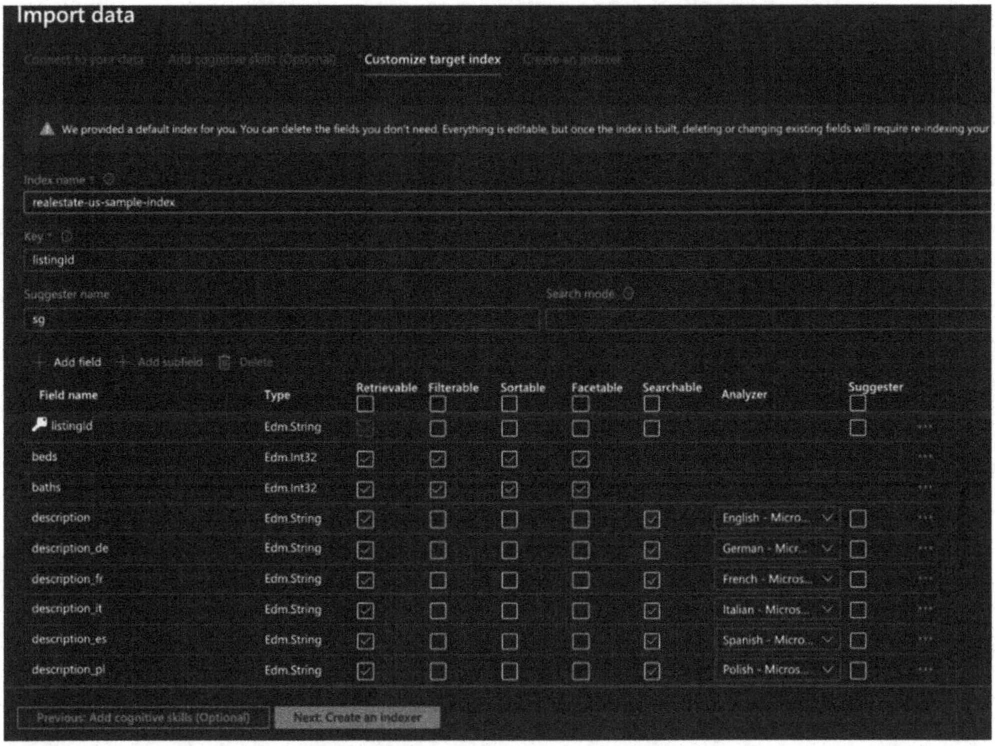

Figure 7-11. *Azure Cognitive Search – Customize target index*

The final step is to create an indexer and define the frequency you would like it to run (see Figure 7-12).

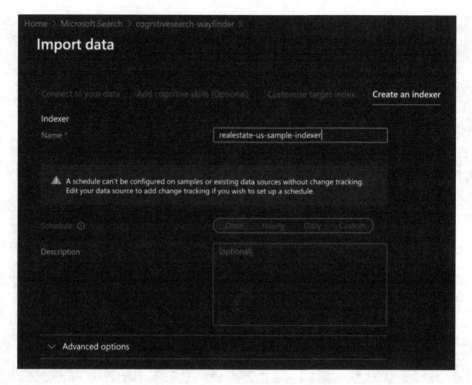

Figure 7-12. Azure Cognitive Search – Create an indexer

Testing Azure Cognitive Search

Once the indexer is created, you can go back to the dashboard and see the progress. You can invoke the service directly from the console. The search explorer provides you with the capability to do it right from the Azure portal, so that you can see the request and the response (result) and make sure it satisfies your business needs. Figure 7-13 demonstrates invoking the service from the search explorer.

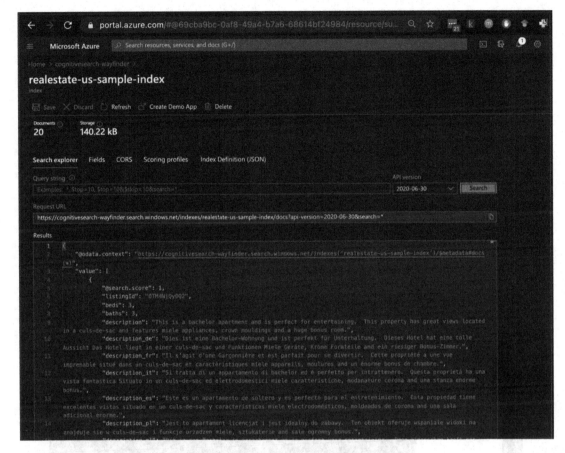

Figure 7-13. *Azure Cognitive Search – invoking the service*

Microsoft has made it really easy for you to consume the service as part of your application. The search explorer tool also comes with a prebuilt HTML and JavaScript-based application, which serves as a getting started guide to build and test the search results in your application. Click **Create Demo App** to continue, and you will see the screen shown in Figure 7-14, which requests you to enable CORS (cross-origin resource sharing).

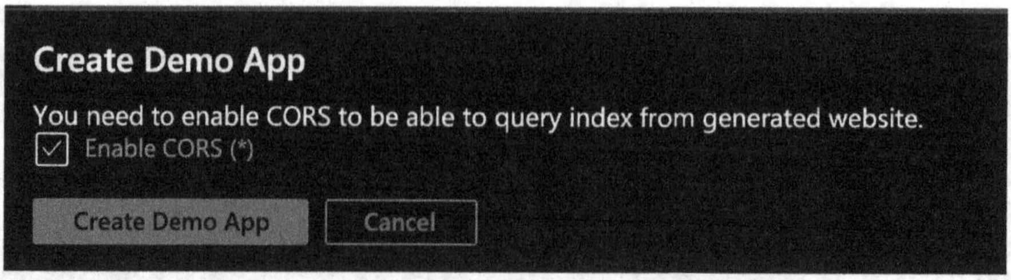

Figure 7-14. *Azure Cognitive Search – enabling CORS*

CORS is an HTTP header-based mechanism, which provides restriction measures for cross-domain invocation. By enabling it, you allow that an HTML file sitting on your desktop can invoke a RESTful API on Azure. In reality (read production), you probably want to be more careful, but this works for the demo app. Now you can start configuring the demo app for results, sidebar, and suggestions (as shown in Figure 7-15).

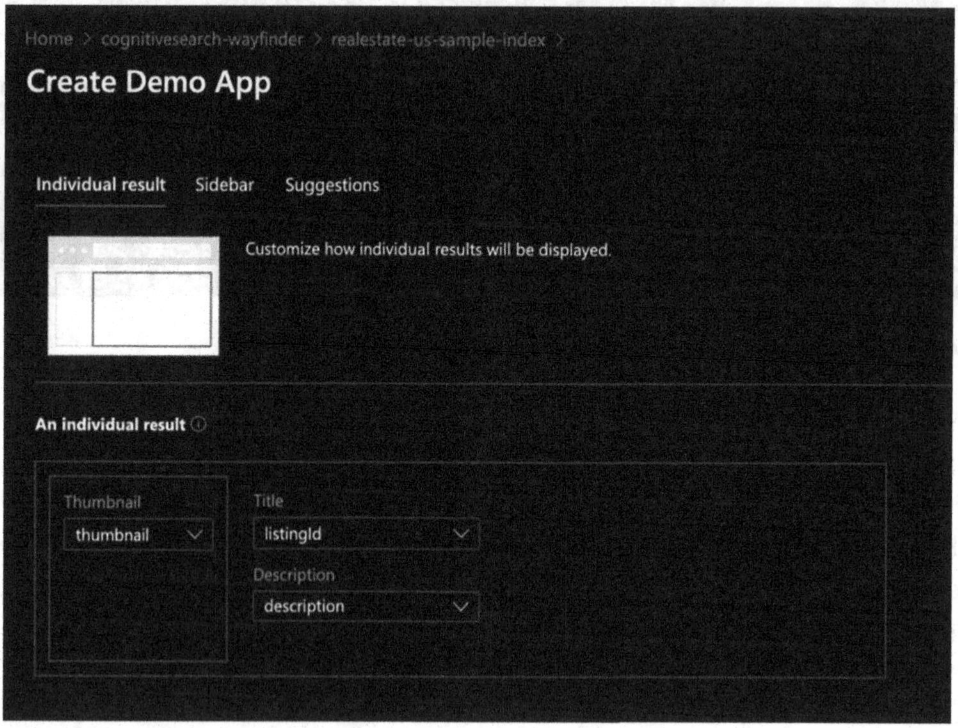

Figure 7-15. *Azure Cognitive Search – Create Demo App*

And voila! We have a search engine for product listing. The HTML file AzSearch. html contains the code required to make the API call and to get the results for search and recommendations. For instance, we search for a home, as shown in Figure 7-16.

Figure 7-16. *Azure Cognitive Search – real estate search application*

The search returns the possible matches, as shown in Figure 7-17.

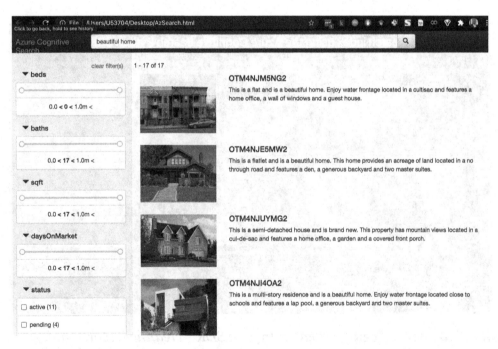

Figure 7-17. *Azure Cognitive Search – search application results*

It is impressive to see how quickly we can get started with a searchable real estate repository. Because the data already exists, the ability to search at lightning speed, with a minimal-code or no-code approach, is empowering for a lot of organizations.

In the next example, we will use Azure Cognitive Search as part of the Jupyter notebook.

Integrating Azure Cognitive Search in Your Notebooks

In Chapter 6, we discussed Jupyter notebook, Anaconda, and the sheer effectiveness of these self-contained, executable Python notebooks, which come in very handy for data scientists. In this example, we will demonstrate how can you create and invoke the Azure Cognitive Search Python SDK.

1. To get started, clone the repository[3] from GitHub (at `https://github.com/Azure-Samples/azure-search-python-samples/tree/master/Quickstart/v11`). Open it up as part of your Jupyter notebook, as shown in Figure 7-18.

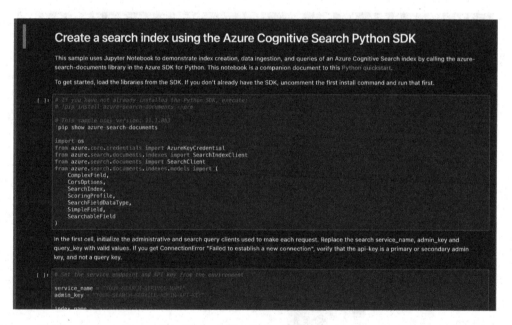

Figure 7-18. *Azure Cognitive Search Python SDK – creating a search index*

---

[3]QuickStart in Python – Jupyter notebook `https://docs.microsoft.com/en-us/samples/azure-samples/azure-search-python-samples/python-sample-quickstart/`

Once opened, uncomment the *azure-search-documents* package to ensure you have the prerequisites, and then run the first cell. Behind the scenes, you loaded the Azure Cognitive Search Sample hotel dataset[4] from JSON files, which consists of 50 hotels in US cities with images, hotel, and room information. See Figure 7-19.

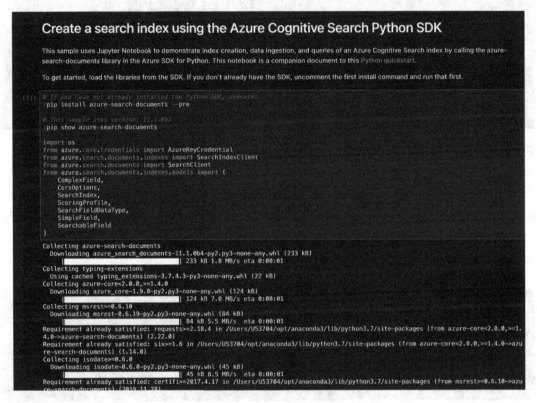

Figure 7-19. *Azure Cognitive Search – installing the prerequisites*

Once completed with the prerequisites, populate the service name and admin key with the information. You will find this information in the dashboard, keys, and endpoint section. The code then creates the SDK client, as shown in Figure 7-20. The client (search client) is used to invoke the service.

---

[4]Azure Cognitive Search Sample Data https://docs.microsoft.com/en-us/samples/ azure-samples/azure-search-sample-data/azure-search-sample-data/

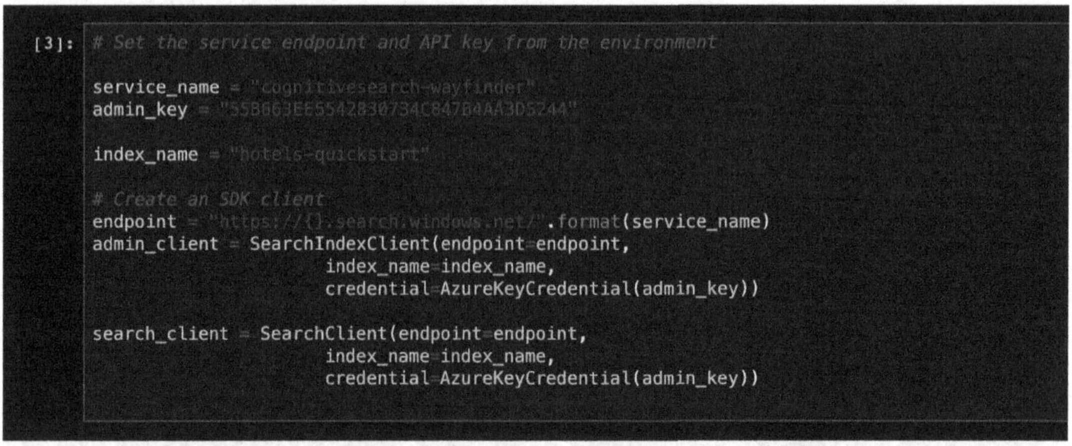

```
[3]:  # Set the service endpoint and API key from the environment

      service_name = "cognitivesearch-wayfinder"
      admin_key = "55B663EE55542830734C847B4AA3D5244"

      index_name = "hotels-quickstart"

      # Create an SDK client
      endpoint = "https://{}.search.windows.net/".format(service_name)
      admin_client = SearchIndexClient(endpoint=endpoint,
                              index_name=index_name,
                              credential=AzureKeyCredential(admin_key))

      search_client = SearchClient(endpoint=endpoint,
                              index_name=index_name,
                              credential=AzureKeyCredential(admin_key))
```

Figure 7-20. *Azure Cognitive Search – setting the parameters*

Now we can check other parameters (like autocomplete) and invoke for incomplete text (like "sa"). You can imagine *sa* being typed, and behind the scenes, you get the result *san Antonio* or *Sarasota* (as shown in Figure 7-21). It then gets filtered by the user. All of these capabilities are built right into the APIs.

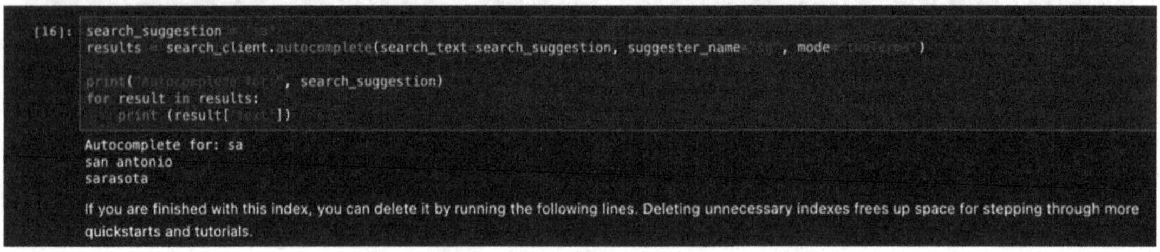

```
[16]:  search_suggestion = "sa"
       results = search_client.autocomplete(search_text=search_suggestion, suggester_name= , mode=  )

       print("Autocomplete for: ", search_suggestion)
       for result in results:
           print (result["text"])

       Autocomplete for: sa
       san antonio
       sarasota

       If you are finished with this index, you can delete it by running the following lines. Deleting unnecessary indexes frees up space for stepping through more
       quickstarts and tutorials.
```

Figure 7-21. *Azure Cognitive Search – autocomplete invocation*

The APIs also allow for invoking the get_document() function to get the information about the specific items, in this case hotels. See the example shown in Figure 7-22.

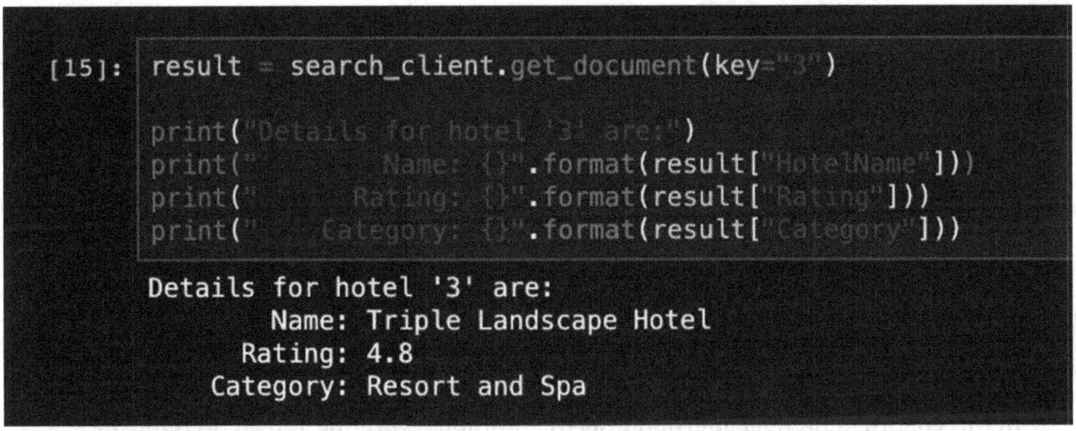
[15]: result = search_client.get_document(key="3")
print("Details for hotel '3' are:")
print("   Name: {}".format(result["HotelName"]))
print("   Rating: {}".format(result["Rating"]))
print("   Category: {}".format(result["Category"]))
Details for hotel '3' are:
        Name: Triple Landscape Hotel
      Rating: 4.8
    Category: Resort and Spa

Figure 7-22. *Azure Cognitive Search - document retrieval*

You can also append to the index and add a list of additional documents, such as hotels (as shown in Figure 7-23).

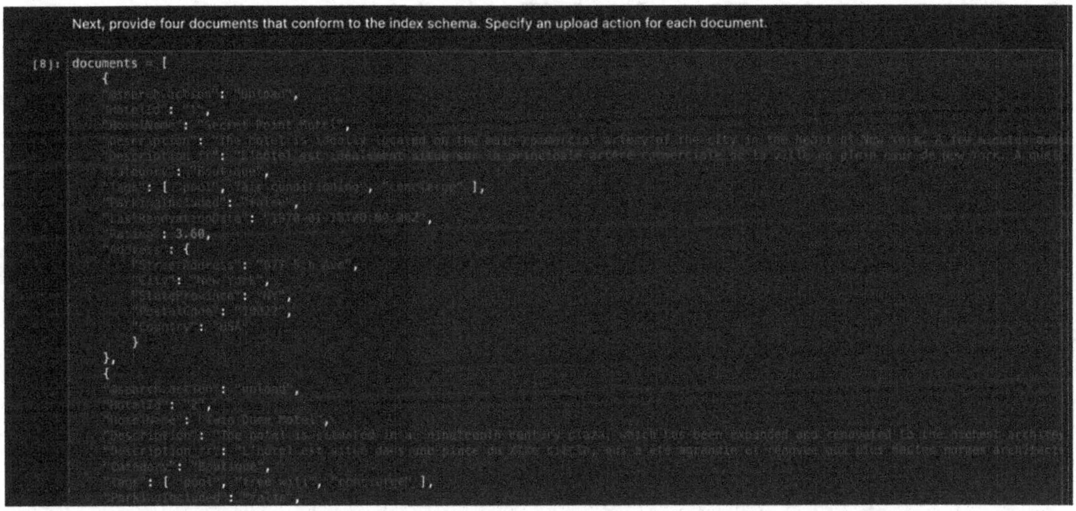
Next, provide four documents that conform to the index schema. Specify an upload action for each document.
[8]: documents = [

Figure 7-23. *Azure Cognitive Search - additional documents (hotels)*

Then, upload the documents to the search client index, as shown in Figure 7-24.

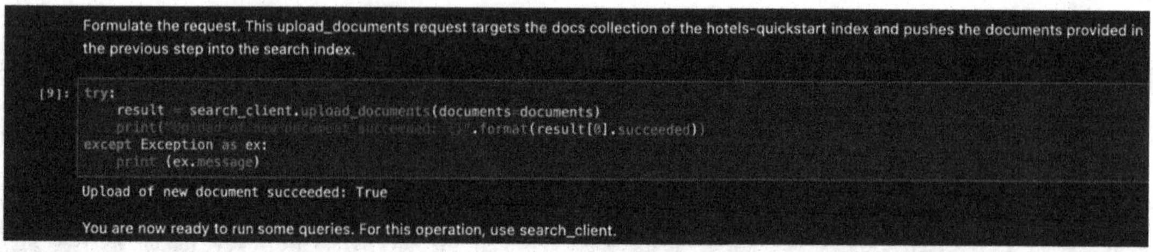

Figure 7-24. *Azure Cognitive Search – uploading additional documents*

Now, you can run the queries on these additional hotels, as these are now part of the search index.

In this section, you have learned how to complete searches, based on datasets like hotels and real estate, which are stored in a structured format (database) or in an unstructured form (documents). Now, we will explore Bing Web Search, to scour billions of pages on the Web for images, videos, news, and more.

Searching with Bing Web Search

Bing Web Search API[5] is a search powerhouse which brings intelligent search to the Web. However, providing its broad reach, it will soon be part of its own Bing Search Services ecosystem. Bing Web Search puts a search engine at your fingertips, by helping you search the web for information stored in web pages, images, videos, and news resources. The information is context and location sensitive, filtered, and free of ads. Spelling correction comes standard, but unlike Azure Cognitive Search, you can tap into the wisdom of crowds by looking into related searches.

Let's start building a Bing Web Search API instance:

1. Access the Bing Web Search API by creating an instance on the
    Azure portal (as shown in Figure 7-25). See `https://portal.`
    `azure.com/#create/microsoft.bingsearch`.

---

[5]Bing Web Search API – `www.microsoft.com/en-us/bing/apis/bing-web-search-api`

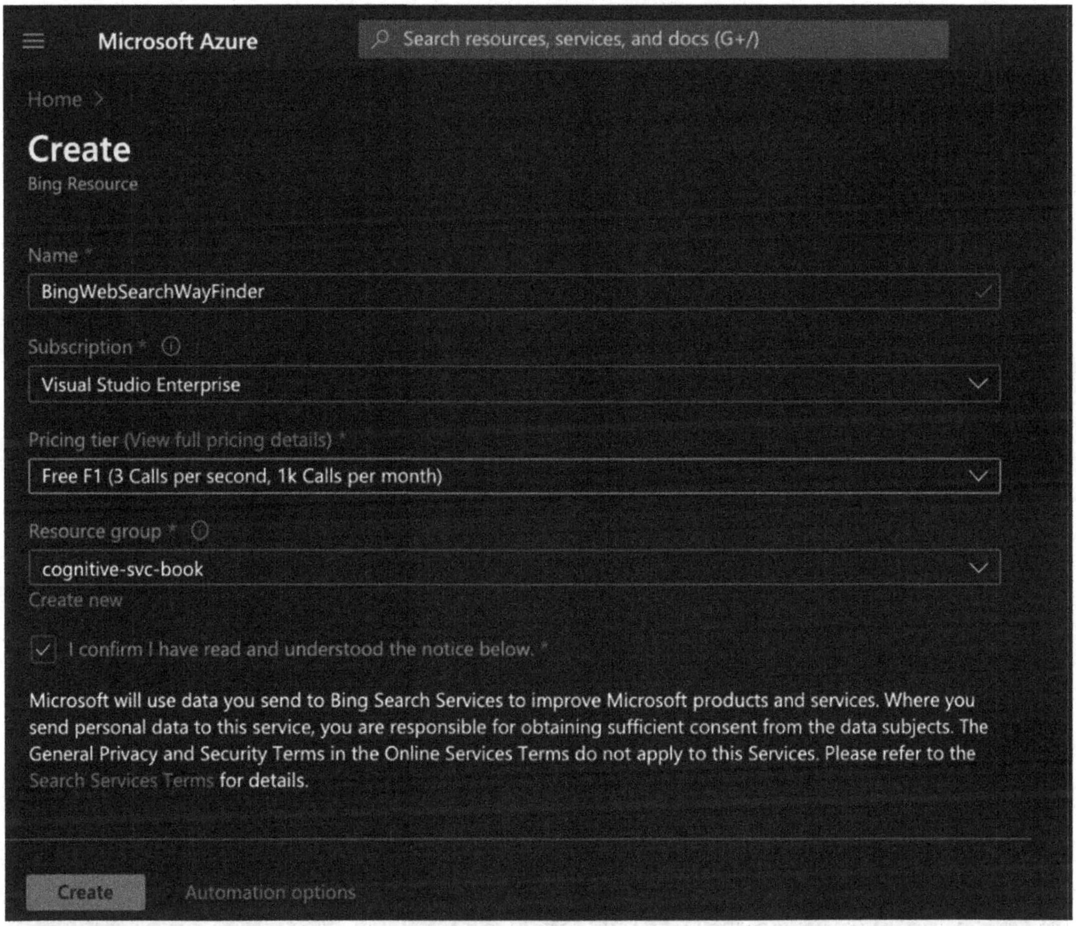

Figure 7-25. *Bing Web Search API – creating an instance*

Once you have completed the information, click **Create**, and it creates a Microsoft
Bing Search instance, as shown in Figure 7-26.

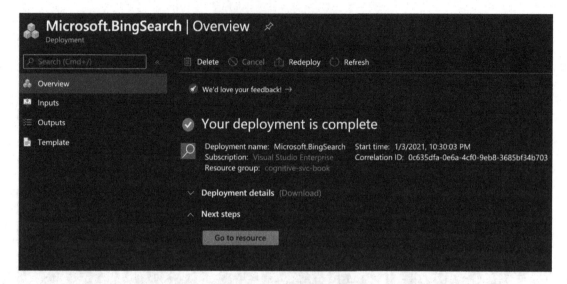

Figure 7-26. *Bing Search API – deployment completed*

Click **Go to resource**, to see the Quick start page (shown in Figure 7-27).

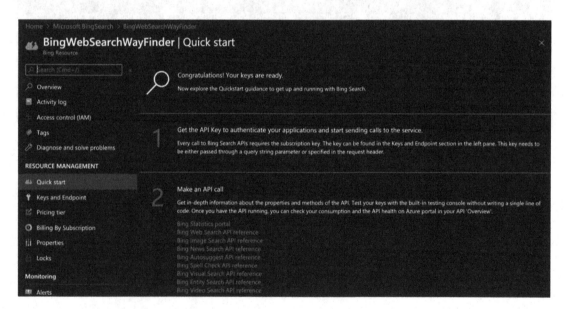

Figure 7-27. *Bing Web Search – Quick start page*

2. In this example, we will complete an image search by using the Bing Image Search APIs. First, we will perform a simple search to retrieve the results, and then we'll use Python Imaging Library to display the results. We are using PyCharm, as shown in Figure 7-28, but feel free to use the IDE of your choice. The code originates from MIT-licensed *cognitive-services-REST-api-samples*[6], but we have made several changes to simplify it and to make it easy to consume. The complete listing is available as part of this book's GitHub repository.

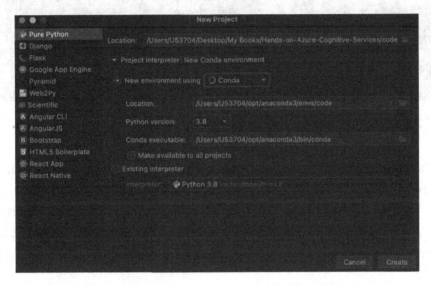

Figure 7-28. *Bing API search in Python – PyCharm project start page*

3. Once the project is created, you will need the keys and endpoints, which can be retrieved from the portal (as shown in Figure 7-29).

---

[6]https://github.com/Azure-Samples/cognitive-services-REST-api-samples/blob/master/python/Search/BingWebSearchv7.py

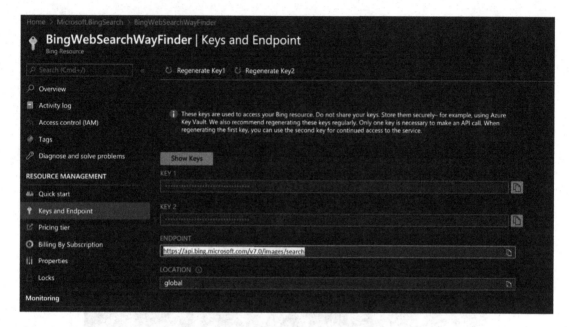

Figure 7-29. *Bing API search – Keys and Endpoint*

Populate the subscriptionKey and endpoint variables with the key and endpoint information. The code in Figure 7-30 is extremely simple and shows you how easy it is to invoke the search service. Aside from subscriptionKey and endpoint as part of the headers, you also define the market via mkt ('en-US') and query ("cat memes"). Run the program to get the results. You can see the response in the output pane shown in Figure 7-30.

Figure 7-30. *Bing Image Search in Python*

Along with individual results, you can open the websearchURL response element in the browser. See Figure 7-31.

Figure 7-31. *Bing Image Search results – GET request in the browser*

4. We will now demonstrate how you can call the API and plot
    the response using Python Imaging Library. The code isn't too
    different. The only additional part is to read the thumbnails and to
    show them on the image canvas.

```python
f, axes = plt.subplots(4, 4)
for i in range(4):
    for j in range(4):
        image_data = requests.get(thumbnail_urls[i + 4 * j])
        image_data.raise_for_status()
        image = Image.open(BytesIO(image_data.content))
        axes[i][j].imshow(image)
        axes[i][j].axis("off")
plt.show()
```

You can see the result in Figure 7-32, where the image is being
drawn on the left pane.

Figure 7-32. *Bing Image Search results in Python*

The complete listing is shown in Listing 7-1.

Listing 7-1. Bing Image Search results in Python

```python
import requests
import matplotlib.pyplot as plt
from PIL import Image
from io import BytesIO

subscriptionKey = '5202a6e4229e4d3491f648d7ebc8aa9c'
endpoint = 'https://api.bing.microsoft.com/v7.0/images/search'

query = "cat memes"

mkt = 'en-US'
params  = {"q": query, "license": "public", "imageType": "photo"}
headers = {'Ocp-Apim-Subscription-Key': subscriptionKey}

try:
    response = requests.get(endpoint, headers=headers, params=params)
    response.raise_for_status()
    search_results = response.json()
    thumbnail_urls = [img["thumbnailUrl"] for img in search_
    results["value"][:16]]

    f, axes = plt.subplots(4, 4)
    for i in range(4):
        for j in range(4):
            image_data = requests.get(thumbnail_urls[i + 4 * j])
            image_data.raise_for_status()
            image = Image.open(BytesIO(image_data.content))
            axes[i][j].imshow(image)
            axes[i][j].axis("off")
    plt.show()

except Exception as ex:
    raise ex
```

The raw JSON response returned by service is shown in Listing 7-2. We have truncated it for brevity and to avoid cuteness overload, with a single result.

Listing 7-2. Bing Image Search results – JSON response

Headers:

```
{'Cache-Control': 'no-cache, no-store, must-revalidate', 'Pragma':
'no-cache', 'Content-Length': '124071', 'Content-Type': 'application/json;
charset=utf-8', 'Expires': '-1', 'P3P': 'CP="NON UNI COM NAV STA LOC CURa DEVa
PSAa PSDa OUR IND"', 'BingAPIs-TraceId': 'C1EEB539F44E4BF8AE333467AFD3A2A9',
'X-MSEdge-ClientID': '324563186B086B31062D6CAE6ABF6A2A', 'X-MSAPI-UserState':
'33d6', 'X-Search-ResponseInfo': 'InternalResponseTime=279,MSDatacenter
=BN2B', 'X-MSEdge-Ref': 'Ref A: C1EEB539F44E4BF8AE333467AFD3A2A9 Ref B:
BLUEDGE0716 Ref C: 2021-01-05T02:47:31Z', 'apim-request-id': '9e4f0b22-8ce6-
421d-a0f0-5cbc1499de3d', 'Strict-Transport-Security': 'max-age=31536000;
includeSubDomains; preload', 'x-content-type-options': 'nosniff',
'CSP-Billing-Usage': 'CognitiveServices.BingSearchV7.Transaction=1', 'Date':
'Tue, 05 Jan 2021 02:47:31 GMT'}
```

JSON Response:

```
{'_type': 'Images',
 'currentOffset': 0,
 'instrumentation': {'_type': 'ResponseInstrumentation'},
 'nextOffset': 42,
 'pivotSuggestions': [{'pivot': 'cat',
                       'suggestions': [{'displayText': 'Baby',
                               'searchLink': 'https://api.bing.microsoft.
com/api/v7/images/search?q=Baby+Memes&tq=%
7b%22pq%22%3a%22cat+memes%22%2c%22qs%22%3a
%5b%7b%22cv%22%3a%22cat%22%2c%22pv%22%3a%2
2cat%22%2c%22hps%22%3atrue%2c%22iqp%22%3af
alse%7d%2c%7b%22cv%22%3a%22memes%22%2c%22p
v%22%3a%22memes%22%2c%22hps%22%3atrue%2c%2
2iqp%22%3afalse%7d%2c%7b%22cv%22%3a%22Baby
%22%2c%22pv%22%3a%22%22%2c%22hps%22%3afals
e%2c%22iqp%22%3atrue%7d%5d%7d',
```

```
'text': 'Baby Memes',
'thumbnail': {'thumbnailUrl':
'https://tse2.mm.bing.net/
th?q=Baby+Memes&pid=Api&mkt=en-
US&adlt=moderate&t=1'},
'webSearchUrl':
'https://www.bing.com/images/searc
h?q=Baby+Memes&tq=%7b%22pq%22%3a%2
2cat+memes%22%2c%22qs%22%3a%5b%7b%
22cv%22%3a%22cat%22%2c%22pv%22%3a%
22cat%22%2c%22hps%22%3atrue%2c%22i
qp%22%3afalse%7d%2c%7b%22cv%22%3a%2
2memes%22%2c%22pv%22%3a%22memes%22%
2c%22hps%22%3atrue%2c%22iqp%22%3afa
lse%7d%2c%7b%22cv%22%3a%22Baby%22%2
c%22pv%22%3a%22%22%2c%22hps%22%3afa
lse%2c%22iqp%22%3atrue%7d%5d%7d&FOR
M=IRQBPS'},
```

. . .

As you would notice in the JSON response, it provides the suggestions, search link for direct access, information about the thumbnail, and header-related information. Your enterprise use case would probably not be searching for cat memes, unless you work for a pet food company's social media department. Whatever it might be, you can easily accomplish it with the help of Bing Search APIs.

Summary and Conclusion

The search is over, at least for now.

In this chapter, you have developed an understanding of the Azure Search ecosystem. You explored the Microsoft Azure Cognitive Search capabilities, and you looked into how Microsoft Bing Search works. You have created sample applications to search real estate and hotels by keywords, and you built an application that searches images on the Web.

You are now ready to go and build your own applications using these technologies. Your use cases will be different, but these fundamentals will help you get started with the search, both within your organization and outside on the Web.

Keep exploring!

References and Further Reading

Azure Cognitive Search
https://azure.microsoft.com/en-us/services/search/
Microsoft Bing Web Search API
www.microsoft.com/en-us/bing/apis/bing-web-search-api
AI enrichment in Azure Cognitive Search
https://docs.microsoft.com/en-us/azure/search/cognitive-search-concept-intro

CHAPTER 8

Deploy and Host Services Using Containers

Well, if you have not heard of containers, you might still be using dial-up Internet! Let's just say that containers are the best thing since sliced bread. Containers enable you to deploy your software in a consistent and repeatable manner, in multiple environments. They are not as nearly as bulky as VMs. Scott Hanselman, Microsoft explainer in chief, has done an excellent video[1] on containers, which I can't recommend enough, especially if you are new to this container ecosystem.

In this chapter, our goal is to accomplish insight on Cognitive Services containers. We will highlight the key containerization features, and we'll demonstrate how to build and deploy an application using Docker. In this chapter, you will learn the following topics:

1. Getting started with Cognitive Services containers

2. Understanding deployment and how to deploy and run a container on Azure Container Instances

3. Understand Docker compose and use it to deploy multiple containers

4. Understanding Azure Kubernetes Service and how to deploy an application to Azure Kubernetes Service

Containerization technologies include Docker, Kubernetes (referred to as K8s), Azure Kubernetes Service (AKS), Azure Container Instances (ACI), and other related topics. These topics are far too complex to be covered exhaustively in this chapter.

---

[1]Docker 101 and how do containers work? www.hanselman.com/blog/docker-101-and-how-do-containers-work

© Ed Price, Adnan Masood, and Gaurav Aroraa 2021
E. Price et al., *Hands-on Azure Cognitive Services*, https://doi.org/10.1007/978-1-4842-7249-7_8

Instead, this chapter will focus on getting you started with Cognitive Services containers, explain their need, and enable you to fully utilize the power of these amazing technologies, should you need them.

Cognitive Services Containers

Azure Cognitive Services are hosted in the highly available Azure cloud data centers and provide excellent SLAs (service-level agreements), and they use the economy of scale to learn, adapt, and improve. Why do we need to box them up and serve it as containers?

Great question, but have you spoken to an auditor? Imagine working in a highly regulated environment, where exposing data outside the firewall requires an act of congress, and rightfully so. Also, containers make a great business use case for disconnected environments, those with low latency requirements, or where the Internet connectivity is either unavailable or inconsistent. With full control on data, and model updates, containers provide an excellent alternative to cloud-based services with high throughput and low latency.

Therefore, to solve security, compliance, and operational issues, Azure Cognitive Services provides containers in the following categories:

- **Vision containers** – OCR, perform spatial analysis, face detection, and form recognition

- **Language containers** – sentiment analysis, key phrase extraction, text language detection, Text Analytics for health, and LUIS (language understanding intelligent service or simply *language understanding*)

- **Speech containers** - speech to text, custom speech to text, text to speech, custom text to speech, neural text to speech, and speech language detection

- **Decision container** – Anomaly Detector

Cognitive computing containers are, interestingly enough, Linux containers. Microsoft is the first provider of pretrained AI in containers with the metering model of billing per transaction. For versioning and other related frequently asked questions, Microsoft maintains an active FAQ section on their production documentation which can be found at https://docs.microsoft.com/en-us/azure/cognitive-services/containers/container-faq.

The availability of these offerings varies. Some of the containers are GA (generally available), while others are in preview or a gated private preview (they require access approvals). The current availability status of each offering, for Azure Cognitive Services containers, can be found at `https://docs.microsoft.com/en-us/azure/cognitive-services/cognitive-services-container-support`.

As part of vision containers, the face detection and form recognition services are unavailable. On June 11, 2020, Microsoft announced *"that it will not sell facial recognition technology to police departments in the United States until strong regulation, grounded in human rights, has been enacted."* Other major cloud providers also followed suit. The Form Recognizer service is available as part of the Azure Cognitive Services offering, but not as a container.

The Cognitive Services containers connect to the Azure data centers to send the metering information, but they do not send customer data to Microsoft. Cognitive Services include most industry standard certifications, which helps you build and deploy healthcare, manufacturing, and financial use cases on the Azure platform. These certifications include ISO 20000-1:2011, ISO 27001:2013, ISO 27017:2015, ISO 27018:2014, and ISO 9001:2015 certification; HIPAA BAA, SOC 1 Type 2, SOC 2 Type 2, and SOC 3 attestation; and PCI DSS Level 1 attestation. Cognitive Services containers, however, do not have any compliance certifications.

Note from Microsoft on Face Recognition[2]   On June 11, 2020, Microsoft announced that it will not sell facial recognition technology to police departments in the United States until strong regulation, grounded in human rights, has been enacted. As such, customers may not use facial recognition features or functionality included in Azure Services, such as Face or Video Indexer, if a customer is, or is allowing use of such services by or for, a police department in the United States.

Like SDLC (software design life cycle), the machine learning development life cycle, data science life cycle, and/or Microsoft team data science process also demands a higher degree of control over the deployment process. This includes versioning,

---

[2]`https://docs.microsoft.com/en-us/azure/cognitive-services/cognitive-services-container-support`

controlling model drift, model decay, and compliance. Containers provide you the required granularity and control to ensure that you have the required governance powers.

Next, let's build some Cognitive Services containers.

Hosting Cognitive Services Containers

Cognitive Services containers have multiple hosting options. You can build a Docker container and host it locally and then connect it with Cognitive Services; you can use Azure Container Instances (ACI), Azure Kubernetes Service (AKS), or deploy a Kubernetes cluster on Azure Stack. These options provide the entire spectrum of container deployment with varying degrees of control and responsibility. With great power comes great responsibility.

Running a Language Service Container

We will start small, with building and hosting a simple, local Docker container:

1. First things first, let's install Docker Desktop. You can download the application from `www.docker.com/products/docker-desktop`. It will allow you to build and run the containers with container runtime. It also provides the Docker CLI, the command line interface that's used to run commands. Go ahead and install it on your platform; we will wait.

2. Our goal in this first example is to run a Cognitive Services container instance locally, connect it with the Text Analytics service for metering and billing, and then invoke the language detection service on the container. We will use a few different containers and services in this chapter. You can find a detailed list of Azure Cognitive Services container image tags and release notes here[3].

---

[3]`https://docs.microsoft.com/en-us/azure/cognitive-services/containers/container-image-tags?tabs=current`

3. Cognitive Services containers are self-sufficient, to run the underlying Cognitive Services. However, they need to be able to access the subscription for metering and billing purposes. Therefore, you would need to create a Text Analytics services account by visiting `https://portal.azure.com/#create/hub`, as shown in Figure 8-1.

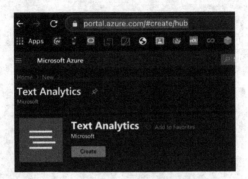

Figure 8-1. *Text Analytics*

4. Complete the required fields (as shown in Figure 8-2), including subscription, location, pricing tier, and resource group. Then, click **Create**. This process has been repeatedly discussed in Chapters 6 and 7 in detail, so refer back if you need to, as we now will skip over some steps for brevity.

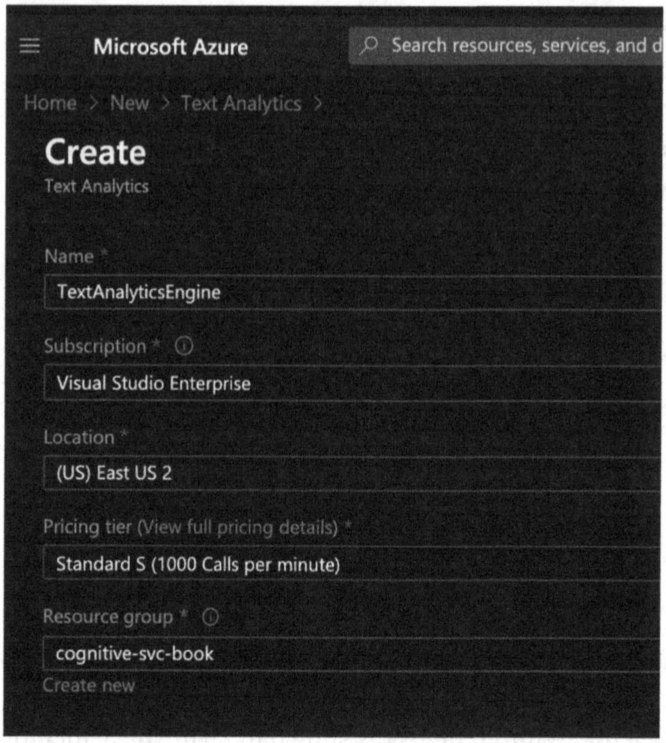

Microsoft Azure

Home > New > Text Analytics >

Create
Text Analytics

Name *

TextAnalyticsEngine

Subscription * ⓘ

Visual Studio Enterprise

Location *

(US) East US 2

Pricing tier (View full pricing details) *

Standard S (1000 Calls per minute)

Resource group * ⓘ

cognitive-svc-book

Create new

Figure 8-2. *Create Text Analytics*

Once the service is validated, created, and deployed, you will see
the screen in Figure 8-3. To continue, click **Go to resource**.

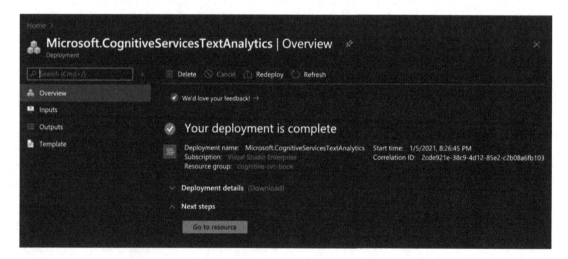

Figure 8-3. *Your deployment is complete*

As shown in Figure 8-4, you will get the key and endpoints from the Text Analytics engine service dashboard, which you can use in the container.

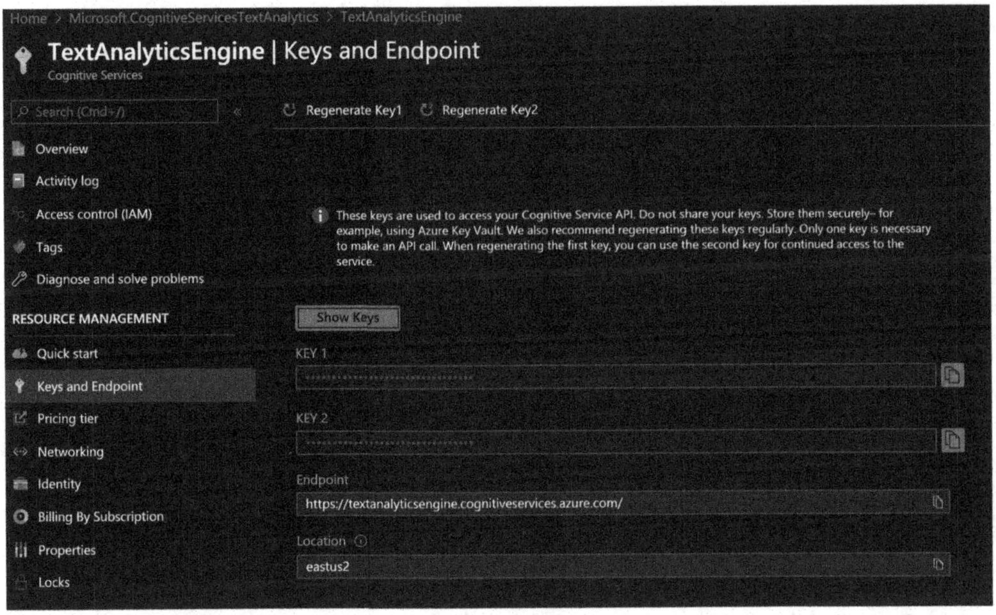

Figure 8-4. Keys and Endpoint

5.   Now that we have the keys to the analytics service for metering, we can proceed with building and running the container. We are running this on a Mac with the configuration shown in Figure 8-5. You likely need to allocate some memory to your container, so make sure you have a decent machine with good CPU and memory to be able to run the container locally. Otherwise, it will be super slow and may not run.

macOS Big Sur
Version 11.1

MacBook Pro (15-inch, 2019)
Processor   2.3 GHz 8-Core Intel Core i9
Memory   32 GB 2400 MHz DDR4

Figure 8-5. macOS Big Sur

The Docker container for language detection is located here[4] on
Docker Hub. It pulls from the Microsoft container repository[5].
In order to start and run the Docker image, you need to run the
following command:

```
docker run --rm -it -p 5000:5000 --memory 8g --cpus 1 \
mcr.microsoft.com/azure-cognitive-services/language \
Eula=accept \
Billing={ENDPOINT_URI} \
ApiKey={API_KEY}
```

You then need to replace the endpoint URL and API key, as shown
in the following. Please use your own keys.

```
docker run --rm -it -p 5000:5000 --memory 8g --cpus 1 \
mcr.microsoft.com/azure-cognitive-services/language \
Eula=accept \
Billing='https://textanalyticsengine.cognitiveservices.azure.com/' \
ApiKey='5e9a5dcd31d9458f8e6f51a488d86399'a
```

Running this code in the terminal will pull down the container
image from the registry. Run it locally, as shown in Figure 8-6.

---

[4]https://hub.docker.com/_/microsoft-azure-cognitive-services-language
[5]Microsoft container repository – Azure Cognitive Services Language Detection Images
mcr.microsoft.com/azure-cognitive-services/language

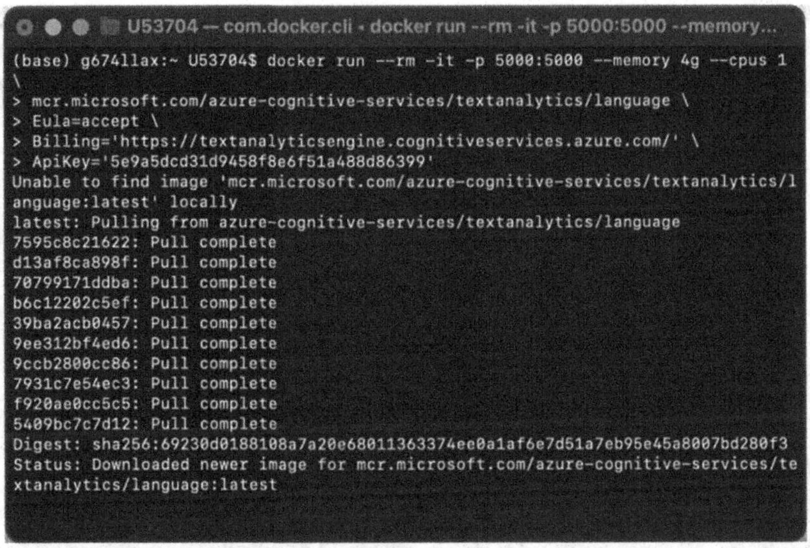

```
U53704 — com.docker.cli · docker run --rm -it -p 5000:5000 --memory...
(base) g674llax:~ U53704$ docker run --rm -it -p 5000:5000 --memory 4g --cpus 1
\
> mcr.microsoft.com/azure-cognitive-services/textanalytics/language \
> Eula=accept \
> Billing='https://textanalyticsengine.cognitiveservices.azure.com/' \
> ApiKey='5e9a5dcd31d9458f8e6f51a488d86399'
Unable to find image 'mcr.microsoft.com/azure-cognitive-services/textanalytics/l
anguage:latest' locally
latest: Pulling from azure-cognitive-services/textanalytics/language
7595c8c21622: Pull complete
d13af8ca898f: Pull complete
70799171ddba: Pull complete
b6c12202c5ef: Pull complete
39ba2acb0457: Pull complete
9ee312bf4ed6: Pull complete
9ccb2800cc86: Pull complete
7931c7e54ec3: Pull complete
f920ae0cc5c5: Pull complete
5409bc7c7d12: Pull complete
Digest: sha256:69230d0188108a7a20e68011363374ee0a1af6e7d51a7eb95e45a8007bd280f3
Status: Downloaded newer image for mcr.microsoft.com/azure-cognitive-services/te
xtanalytics/language:latest
```

Figure 8-6. Pulling down the container image

You can see the list of images available in the Docker Desktop console, as shown in Figure 8-7.

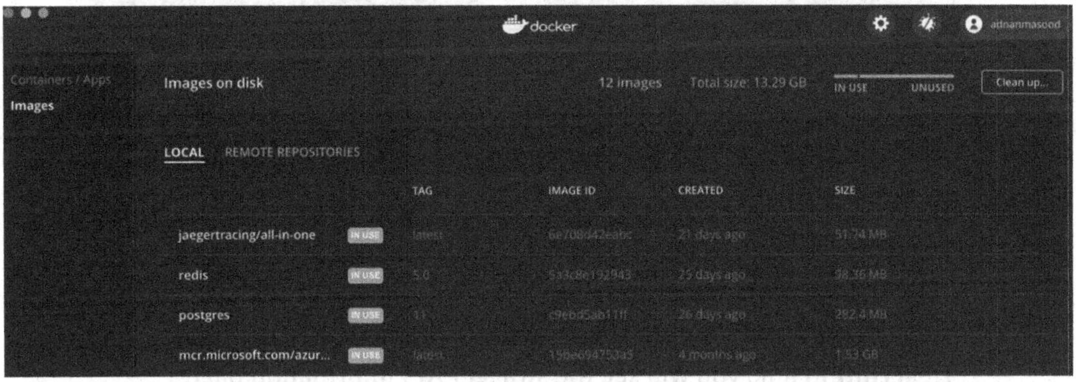

Figure 8-7. Images on disk

There is no place like 127.0.0.1. So once the image is downloaded, you can visit the localhost:5000, as it's perfectly safe during the COVID lockdowns. The URL and port were specified in the command, if you didn't notice (be sure to copy and paste!). Also note that your Azure Cognitive Services container is up and running, as shown in Figure 8-8. If you would like to run multiple containers, change the port numbers.

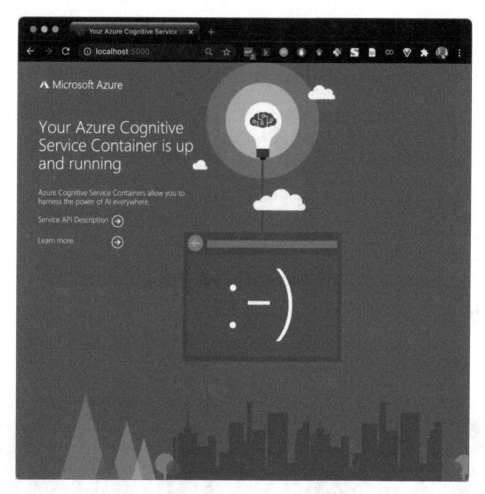

Figure 8-8. *Azure Cognitive Service container confirmation*

The language container (and therefore the language detection service) is up and running. If you click the **Service API Description** link, you will see the swagger API information page (shown in Figure 8-9), where you can invoke the services.

Figure 8-9. *Language Detection Cognitive Service API*

There are different versions of the services available on the container. You can click any of these links and invoke the service with a pre-created request. See Figure 8-10.

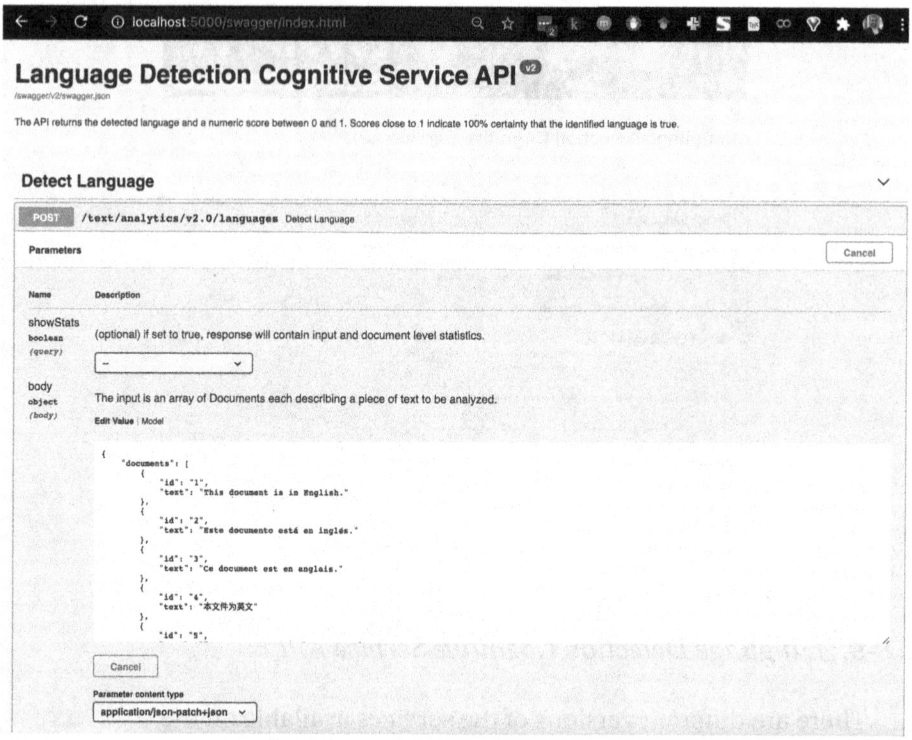

Figure 8-10. *A pre-created request*

Click **Execute**, and then you'll see the response. Next, you will also make the cURL[6] request, in case you feel like invoking it from the terminal. See the following code:

```
curl -X POST "http://localhost:5000/text/analytics/
v2.0/languages" -H "accept: application/json"
-H "Content-Type: application/json-patch+json"
-d "{ \"documents\": [ { \"id\": \"1\", \"text\":
\"This document is in English.\" }, { \"id\": \"2\",
\"text\": \"Este documento está en inglés.\" },
{ \"id\": \"3\", \"text\": \"Ce document est en
anglais.\" }, { \"id\": \"4\", \"text\":
\"本文件为英文\" }, { \"id\": \"5\", \"text\":
\"Этот документ на английском языке.\" } ]}"
```

---

[6]https://curl.se/docs/manpage.html?data1=dwnmop

Here, we executed the code. You can see the response detecting the languages correctly in Figure 8-11.

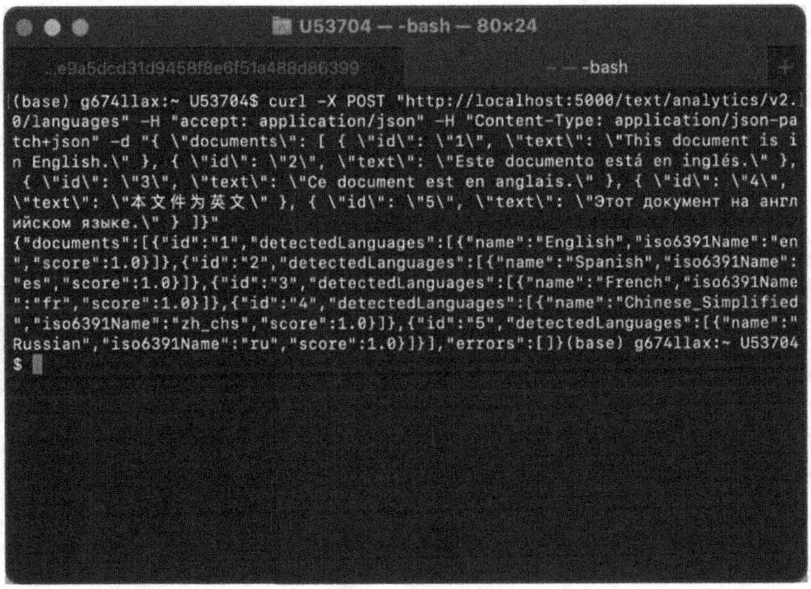

Figure 8-11. *Detecting the languages*

The docker command to run the container would be

docker run --rm -it -p 5000:5000 --memory 8g --cpus 1 mcr.microsoft.com/azure-cognitive-services/textanalytics/language:1.1.012840001-amd64 Eula=accept   Billing="https://<resource_name>.cognitiveservices.azure.com/" ApiKey="<subscription_key>"

You can also invoke it directly from the web page as well. The screen in Figure 8-12 shows the response you would get from the container, via the web page.

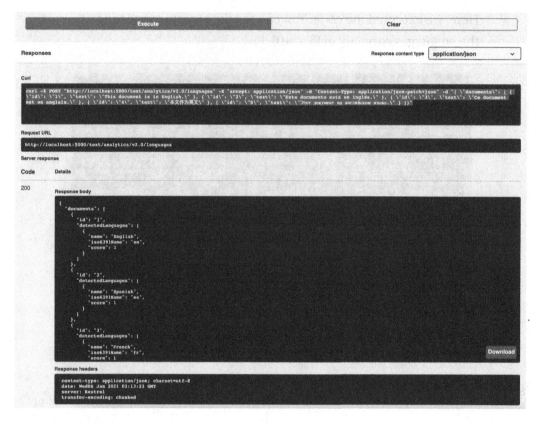

Figure 8-12. The response from the container

In this example, we pulled a Cognitive Services Docker container, ran it while linking it with the metering language service, and we invoked the results via the webUI and cURL interfaces. In the next example, let's try to do this with the Anomaly Detector service, but this time, we'll use Jupyter notebook.

Running an Anomaly Detector Service Container

1. Similar to the language detection container, Anomaly Detectors are hosting on the Docker Hub[7], as well as the Microsoft container repository at mcr.microsoft.com/azure-cognitive-services/decision/anomaly-detector.

---

[7]https://hub.docker.com/_/microsoft-azure-cognitive-services-decision-anomaly-detector

2. Create an instance of the Anomaly Detector service to get the endpoint and keys for metering (see Figure 8-13). You can refer to Chapter 6 for the detailed steps of creating an anomaly detection service, but it's pretty much identical to what we have done earlier with the language service. Once an instance is created, we would use it in the Docker run command.

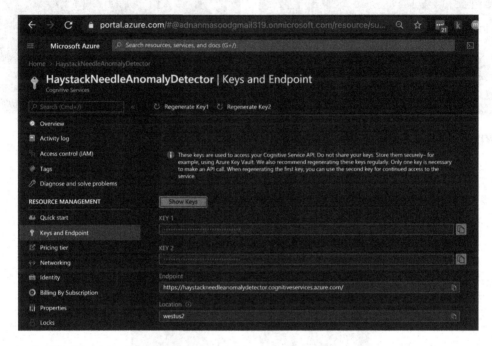

Figure 8-13. *Keys and Endpoint*

3. Run the following command to create an Anomaly Detector Docker container. This will pull down the latest container and connect the service to the Azure portal for billing.

```
docker run --rm -it -p 5000:5000 --memory 4g --cpus 1 \
mcr.microsoft.com/azure-cognitive-services/decision/anomaly-
detector:latest \
Eula=accept \
Billing='https://haystackneedleanomalydetector.cognitiveservices.
azure.com/' \
ApiKey='83c02850ba4b402f819c0f2f87f71b86'
```

You would then see the screenshots shown in Figure 8-14, as the application starts serving on the `http://localhost:5000` URL.

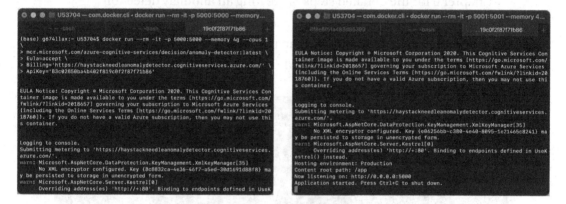

Figure 8-14. *The application starts serving*

4. As shown in the language detection example earlier, you can access the Cognitive Services container at `http://localhost:5000` and see the welcome screen shown in Figure 8-15. This screen is consistent across the Cognitive Services.

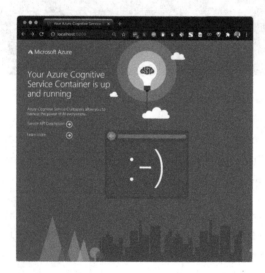

Figure 8-15. *The welcome screen*

Next, you will see the Swagger page for the APIs and invoke them directly from the UI, or via an application (see Figure 8-16).

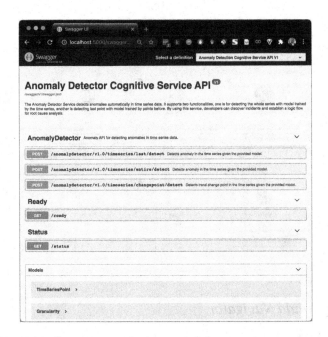

Figure 8-16. *The Swagger page*

5. For this example, we will run a local instance of Jupyter notebook
    using Anaconda, and we'll run the sample notebook we used in
    Chapter 6. The repository can be cloned from `https://github.`
    `com/Azure-Samples/AnomalyDetector`. Start the JupyterLab from
    Anaconda Navigator (as shown in Figure 8-17), and then navigate
    to the point anomaly detection notebook.

Figure 8-17. *Anaconda Navigator*

Open the notebook, and then populate the API key and endpoint with the API key from the anomaly detection service and with the URL http://localhost:8888 to the local anomaly detection container instance. See Figure 8-18.

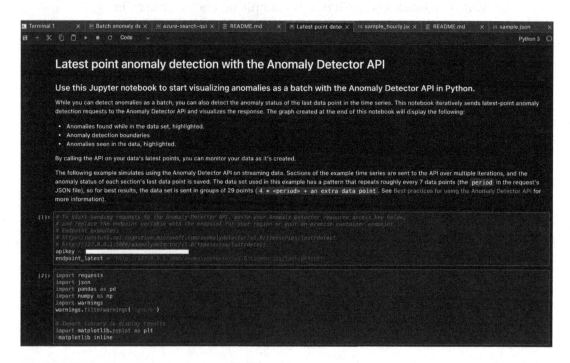

Figure 8-18. *Latest point anomaly detection*

The first cell should look like the cell in Figure 8-19.

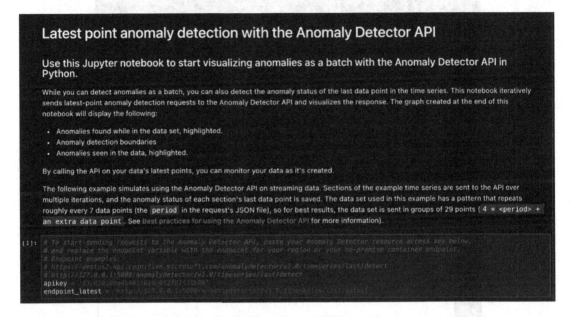

Figure 8-19. *Anomaly Detector resource access key*

Now, when you invoke the further cells to run the example, you will see similar results as shown in Chapter 6. See Figure 8-20. The difference being that the code is now running locally on your machine via the Cognitive Services container.

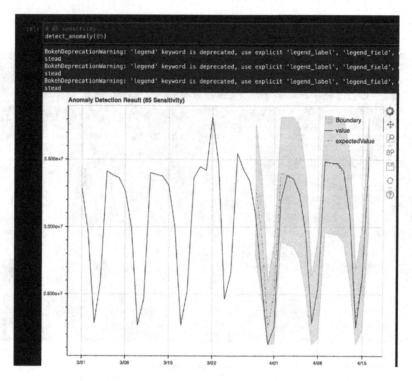

Figure 8-20. *Anomaly Detection Result*

With these two examples, you saw how easy it is to run Cognitive Services containers locally, using Docker. Now we will use the cloud-based approaches of ACI and AKS to host these containers.

Working with Azure Container Instances

"It works on my machine" is a great goal, but you might want to achieve reproducibility, performance, simplicity, and speed, without having the burden of a full orchestration platform (such as Kubernetes and Mesos). Azure Container Instances (ACI) provides a good alternative. ACI is a quick and lightweight way of running containers in the Azure cloud, and you can manage your own container registry with Azure Container Registry (ACR).

"Doesn't this defeat the purpose?" you may ask. The reason for having containers was to have control and not deploy things on the cloud. Now we suggest that you deploy the container itself on the cloud? That does not make any sense. Let us explain.

Besides compliance and data protection, one of the reasons for containerization is control. If your organizational or regional policy allows you to only host, keep, and process your customer data in the specific region, and you want to control your deployment process, then running your containers in ACI might be a good alternative to managing your own data center. With the Microsoft cloud platform, you have choices – you can bring your own hardware, use lightweight ACI, or go with an orchestration platform like AKS. Let's get started, with ACI:

1. Go to the Azure portal and search for "ACI" or "Container instances," as shown in Figure 8-21. Click **Container instances**.

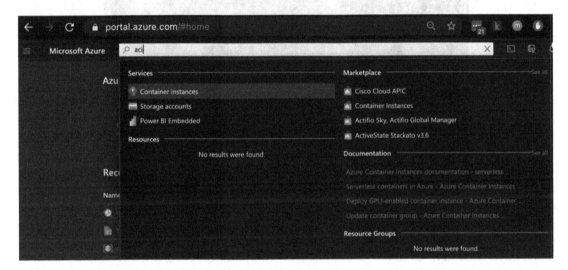

Figure 8-21. *Searching for ACI*

2. Create the container instance, as shown in Figure 8-22. Populate the required fields, and then select the image link. In this case, we select the Microsoft Text Analytics image from MCR.

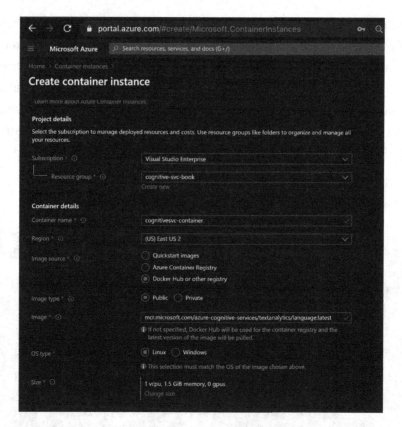

Figure 8-22. *Create container instance*

Before proceeding, you should review the container requirements and recommendations[8], and then make sure that the configuration of your container instances matches the configuration that Microsoft recommends. The current specification is listed in Figure 8-23, but you should verify the necessary specification, as they may periodically change. We now live in the era of a billion parameter models.

_____

[8]https://docs.microsoft.com/en-us/azure/cognitive-services/text-analytics/how-tos/text-analytics-how-to-install-containers?tabs=language

	Minimum host specs	Recommended host specs	Minimum TPS	Maximum TPS
Language detection, key phrase extraction	1 core, 2GB memory	1 core, 4GB memory	15	30
Sentiment Analysis v3	1 core, 2GB memory	4 cores, 8GB memory	15	30
Text Analytics for health - 1 document/request	4 core, 10GB memory	6 core, 12GB memory	15	30
Text Analytics for health - 10 documents/request	6 core, 16GB memory	8 core, 20GB memory	15	30

Figure 8-23. *The container requirements*

In our case, we will use the language detection container, and we will modify the specs by clicking **Change size**. You will then see the Change container configuration screen, as shown in Figure 8-24. Update the number of CPU cores and the memory required.

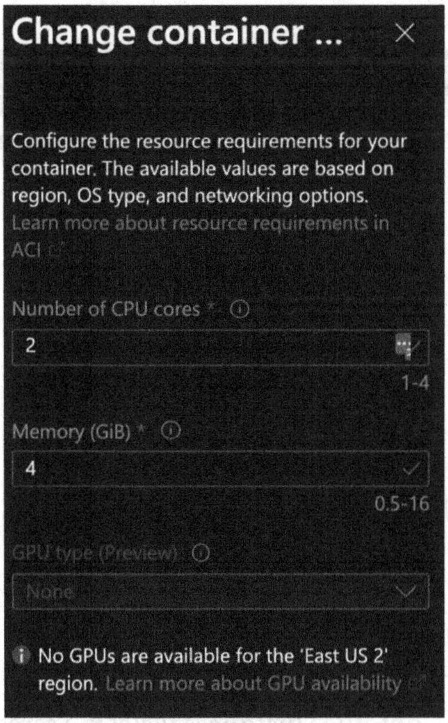

Figure 8-24. *Configuring the resource requirements*

Now you can proceed and create the container. Once the deployment is complete, you would see screen shown in Figure 8-25.

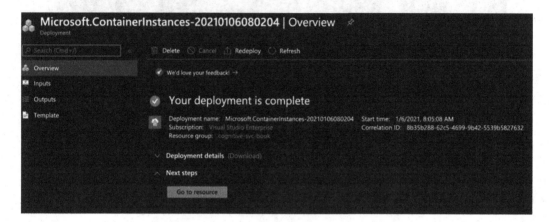

Figure 8-25. *A completed deployment*

Since this container is deployed in the cloud via ACI, it has a public IP (in our case, it is 52.146.70.12). You can access the IP from your browser to see the welcome screen shown in Figure 8-26.

Figure 8-26. *The Azure Container Instances welcome screen*

Now that the instance is ready, you can access the API, invoke the service, and so on. We will see this in detail shortly.

There are three different ways to build and deploy ACI and AKS containers – via the UI, Azure Cloud Shell, and terminal. In the example earlier, you saw how we can accomplish the task of creating and running a container via the UI. Now let's switch to the command line interface (CLI). Azure Cloud Shell is a great way to perform command line work in the browser. You can access it from shell.azure.com, which will ask you to choose your flavor of cloud shell. See Figure 8-27.

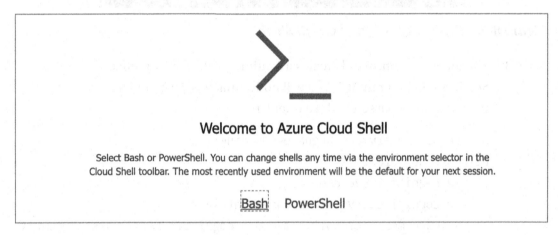

Welcome to Azure Cloud Shell

Select Bash or PowerShell. You can change shells any time via the environment selector in the Cloud Shell toolbar. The most recently used environment will be the default for your next session.

Bash   PowerShell

Figure 8-27. *Selecting a scripting language*

Upon selection of your scripting language, you will see the all-powerful terminal we all love, as shown in Figure 8-28.

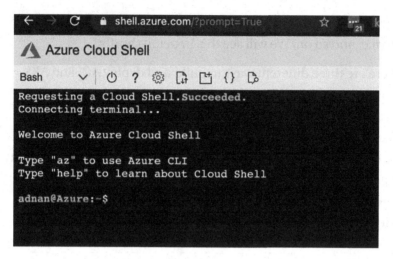

Figure 8-28. *Opening the Azure Cloud Shell*

3. The following command creates a sentiment analysis Cognitive
   Services container in ACI. You will notice that it is very similar to
   the parameters passed in the earlier UI.

```
aci=cognitiveservice-language-container
az container create \
    -g cognitive-svc-book \
    -n cognitiveservice-language-container \
    --image mcr.microsoft.com/azure-cognitive-services/sentiment \
    -e Eula=accept Billing='https://textanalyticsengine.
    cognitiveservices.azure.com/text/analytics/v2.0'
    ApiKey='5e9a5dcd31d9458f8e6f51a488d86399'  \
    --ports 5000 \
    --cpu 4 \
    --memory 16 \
    --ip-address public
```

4. It is important to note that you can run this code on the Azure CLI[9]
   as well, as shown side by side in Figure 8-29. You would need Azure
   CLI installed. In this case, we will only use the Azure Cloud Shell.

---

[9]https://docs.microsoft.com/en-us/cli/azure/install-azure-cli

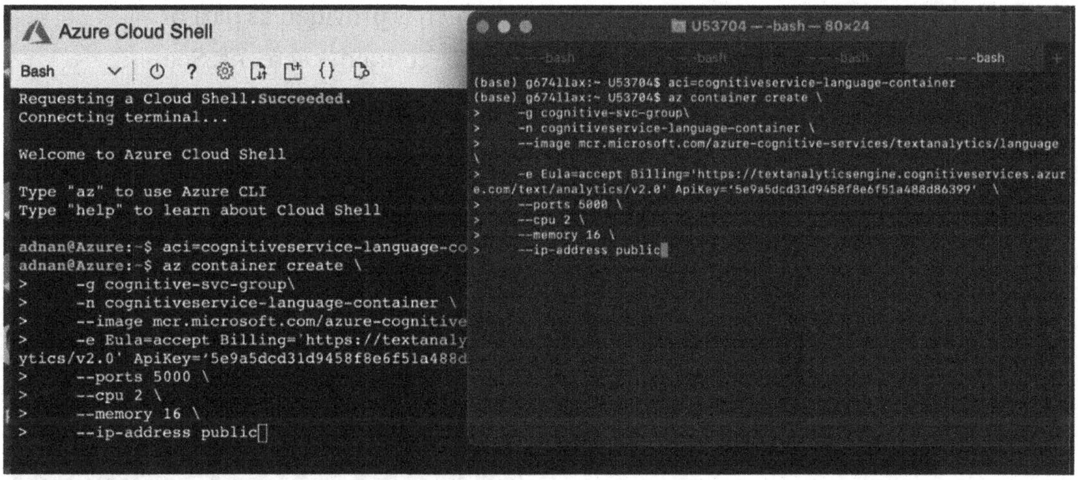

Figure 8-29. *The Azure Cloud Shell and CLI*

As the command runs in the cloud shell, it creates the instance and makes it available via public IP. You will see the completion response, as shown in Figure 8-30.

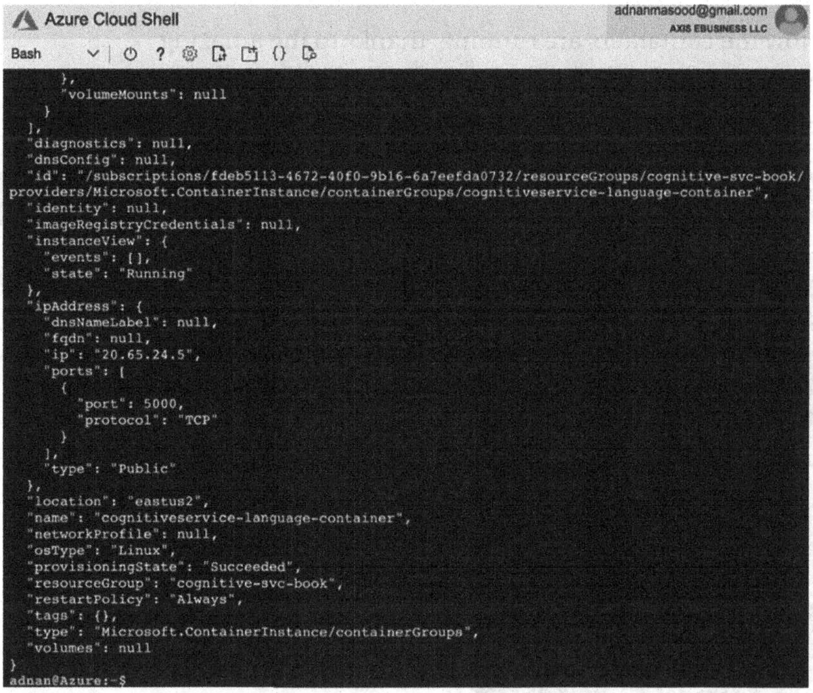

Figure 8-30. *The completion response*

And now you can access the public IP, which is provided as part of a response to access the much sought-after container screens. See Figure 8-31.

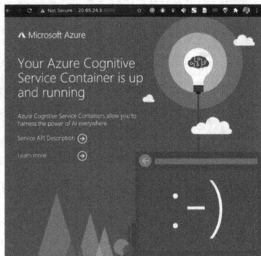

Figure 8-31. *The container welcome screens*

5. Now the containers are running, thanks to the amazing infrastructure and CLI support that's provided by Azure. We can easily invoke the sentiment analysis service.

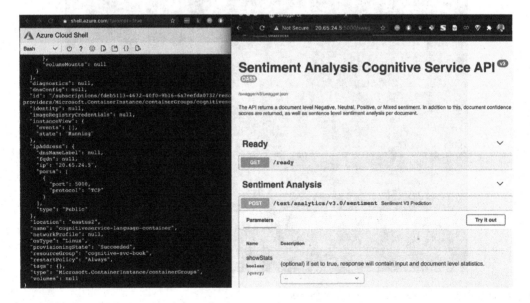

Figure 8-32. *The sentiment analysis service*

6. The following code represents the simple request to get the
    sentiment data:

```json
{
  "documents": [
    {
      "language": "en",
      "id": "1-en",
      "text": "Hello world. This is some input text
      that I love."
    },
    {
      "language": "en",
      "id": "2-en",
      "text": "It's incredibly sunny outside! I'm so happy."
    },
    {
      "language": "en",
      "id": "3-en",
      "text": "Pike place market is my favorite Seattle
      attraction."
    }
  ]
}
```

Once we invoke it via your browser, you will see the response, as
shown in Figure 8-33.

Figure 8-33. *The browser response*

The request body drop-down should be changed to application/json; the default application/json-patch+json will display unsupported content type error.

The JSON looks like the following, which we truncated to one document for brevity:

```
{
  "statistics": {
    "documentsCount": 3,
    "validDocumentsCount": 3,
    "erroneousDocumentsCount": 0,
    "transactionsCount": 3
  },
```

```
"documents": [
  {
    "id": "1-en",
    "sentiment": "positive",
    "confidenceScores": {
      "positive": 1,
      "neutral": 0,
      "negative": 0
    },
    "sentences": [
      ....
  "errors": [],
  "modelVersion": "2019-10-01"
}
```

And there you have it – the creation of an ACI container on Azure, both via a command line and web UI, and with invocation in the cloud.

Now we will review the last approach, Azure Kubernetes Service (AKS), and we'll see how to deploy an Azure Cognitive Services Text Analytics container image to Azure Kubernetes Service.

Deploying a Cognitive Services Container with Azure Kubernetes Service

ACI helps you deploy a container, but what if you have to deploy and manage a bunch of these? Kubernetes is an open source orchestration engine that helps you build, deploy, and manage these containers at scale. Azure Kubernetes Service (AKS) is a managed Kubernetes service on Azure cloud, which, as far as managed services go, is pretty awesome since it's free. Yes, you don't have to pay for AKS, only for the nodes you use. You can create AKS clusters via web UI, CLI, or by using templates.

Let's start with building the Kubernetes Service. As always, portal.azure.com is our one-stop shop. Search for "Kubernetes Service," and then click **Create**. See Figure 8-34.

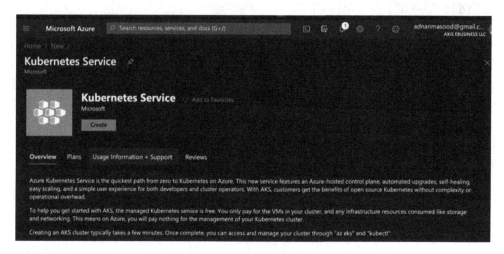

Figure 8-34. *Kubernetes Service in the Azure portal*

Fill out the Create Kubernetes cluster page with the default settings, as shown in Figure 8-35. In the cluster, you have options to customize node pools, authentication, advance networking settings, and integrations. See Figure 8-35. Details about individual parameters, and their possible settings, can be found here[10].

---

[10]https://docs.microsoft.com/en-us/azure/aks/kubernetes-walkthrough-portal

Figure 8-35. *Create Kubernetes cluster*

Click **Review + create** to proceed, and you will see the screen shown in Figure 8-36, once the cluster is deployed (which can take some time).

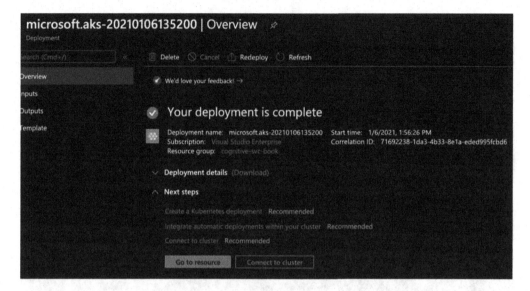

Figure 8-36. *The cluster is deployed*

Once the cluster is deployed, you will want to sign in to the AKS cluster. Use the following command to sign into the cluster:

```
az aks get-credentials --name cognitive-svc-k8 --resource-group cognitive-svc-book
```

You can run this command in the terminal, with Azure CLI installed, and see the output shown in Figure 8-37. Make sure you replace your own resource group and AKS cluster name in the command.

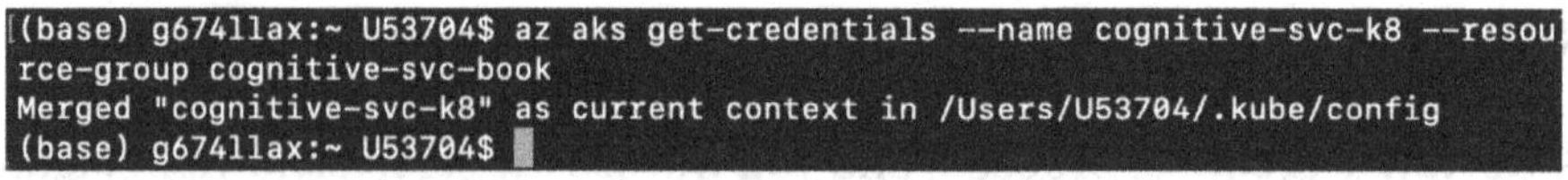

```
(base) g6741lax:~ U53704$ az aks get-credentials --name cognitive-svc-k8 --resou
rce-group cognitive-svc-book
Merged "cognitive-svc-k8" as current context in /Users/U53704/.kube/config
(base) g6741lax:~ U53704$
```

Figure 8-37. *Cluster sign-in output*

The cluster is successfully created, but it does not include any container nodes. So, let's fix that by creating a key phrase Cognitive Services container to the K8s cluster. To accomplish adding containers to the cluster, we will need to add this information in the

K8s config file. Kubernetes uses YAML for its configuration settings. (YAML originally stood for *Yet Another Markup Language* as a silly name, but it was changed to *YAML Ain't Markup Language* to show that it is intended to be used for data-oriented purposes, rather than document markup.)

We will use Visual Studio 2019 for Mac to edit it (as shown in Figure 8-38), but feel free to use the editor of your choice.

Figure 8-38. *Visual Studio 2019 for Mac*

The YAML file (pronounced *yamel*, like *camel*) is fairly simple, as shown in Figure 8-39. Here, we define the container name, the image path, the ports, what kind of resources to allocate, the billing URL, and the API key – similar to what was done earlier for creating an ACI container.

```yaml
1   apiVersion: apps/v1
2   kind: Deployment
3   metadata:
4     name: keyphrase
5   spec:
6     selector:
7       matchLabels:
8         app: keyphrase-app
9     template:
10      metadata:
11        labels:
12          app: keyphrase-app
13      spec:
14        containers:
15        - name: keyphrase
16          image: mcr.microsoft.com/azure-cognitive-services/keyphrase
17          ports:
18          - containerPort: 5000
19          resources:
20            requests:
21              memory: 2Gi
22              cpu: 1
23            limits:
24              memory: 4Gi
25              cpu: 1
26          env:
27          - name: EULA
28            value: "accept"
29          - name: billing
30            value: 'https://textanalyticsengine.cognitiveservices.azure.com/'
31          - name: apikey
32            value: '5e9a5dcd31d9458f8e6f51a488d86399'
33
34  ---
35  apiVersion: v1
36  kind: Service
37  metadata:
38    name: keyphrase
39  spec:
40    type: LoadBalancer
41    ports:
42    - port: 5000
43    selector:
44      app: keyphrase-app
```

Figure 8-39. *The YAML file*

You can now apply this YAML to the container, by using the following command:

```
kubectl apply -f keyphrase.yaml
```

The command line will show the results from Figure 8-40, as the deployment container is created.

```
deployment.apps "keyphrase" created
service "keyphrase" created
```

Figure 8-40. *The deployment container is created*

Now you can verify that the pod (the smallest deployable unit, in this case the Cognitive Service container and associated services and dependencies) was deployed and the services are running, as shown in Figure 8-41.

Figure 8-41. *Verifying the deployment*

As the earlier screenshot shows, the container is now accessible via a public IP address, http://20.75.16.10:5000/, and it can be invoked to gain access to the Cognitive Services container and associated APIs. After connecting to the cluster, you can run a variety of different commands, as shown in Figure 8-42.

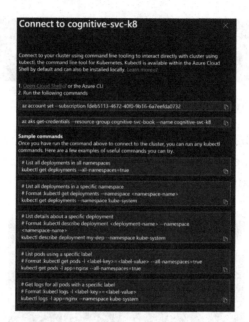

Figure 8-42. *Different commands you can use*

These commands, and the logs screen shown in Figure 8-43, show the amount of control you have over your container management with K8s, as compared to a single individual ACI deployment.

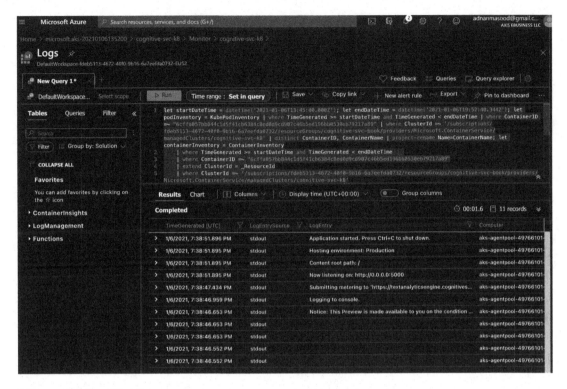

Figure 8-43. *The Logs screen*

Summary and Conclusion

Containers are contained, barely. The world of containers is as big as the thriving discipline of DevOps, and it's extremely difficult to cover it in a short chapter. We tried our best to help you develop an understanding of the Cognitive Services container ecosystem. We explored Docker, ACI, and AKS, and we looked into how the containers work in different settings. You then created samples, which call the self-contained code from containers, without calling the Cognitive Services directly.

You now have the fundamentals to start exploring and to use these technologies for your own business use cases. Your regulatory, compliance, infrastructure, and operational use cases will be different, but these fundamentals will help you get started with containerization within your organization.

Keep containerization!

References and Further Reading

Azure Cognitive Services containers

https://docs.microsoft.com/azure/cognitive-services/cognitive-services-container-support

Install and run Text Analytics containers

https://docs.microsoft.com/azure/cognitive-services/text-analytics/how-tos/text-analytics-how-to-install-containers

Deploy a Text Analytics container to Azure Kubernetes Service

https://docs.microsoft.com/azure/cognitive-services/text-analytics/how-tos/text-analytics-how-to-use-kubernetes-service

Deploy and run container on Azure Container Instances

https://docs.microsoft.com/azure/cognitive-services/containers/azure-container-instance-recipe

Azure Cognitive Services container image tags and release notes

https://docs.microsoft.com/azure/cognitive-services/containers/container-image-tags

CHAPTER 9

Azure Bot Services

Human: What do we want?

> Machine: Context-aware NLP bots

> Human: When do we want it?

> Machine: When do we want what?

Natural language processing (NLP) is a core element of a bot service. Unlike our joke, we want to develop artificial intelligence (AI) in our bots, to understand the context of a conversation. Bots are conversational user interfaces that allow you to communicate with the machines in different modalities, such as text and voice. Azure Bot Services is a cloud based, managed, integrated environment to empower bot development, hosting, and management. Azure Bot Services allow you to focus on what's important, the crux of conversational interface, while it handles the infrastructure and associated toolset. Powered by the comprehensive Bot Framework, Bot Services comes with prebuilt templates for you to get started, providing a browser-based interface to make bot development as easy as possible.

In this chapter, we will provide insights on Bot Services by creating the COVID-19 QnA Bot. You will learn the following topics:

1. Understanding Azure Bot Services

2. Creating a COVID-19 bot using Azure Bot Services

3. Using the Azure Bot Builder SDK[1]

---

[1]Bot Builder SDK is available at `https://docs.microsoft.com/azure/bot-service/dotnet/bot-builder-dotnet-sdk-quickstart`

© Ed Price, Adnan Masood, and Gaurav Aroraa 2021
E. Price et al., *Hands-on Azure Cognitive Services*, https://doi.org/10.1007/978-1-4842-7249-7_9

The Azure Bot Services Ecosystem

The Microsoft Bot ecosystem comprises of a few salient offerings – this includes Microsoft Bot Framework, Azure Bot Services, Language Understanding (LUIS), and Azure Health Bot (conversational AI for healthcare).

Bot Framework has gone through a few iterations. It provides an SDK and a comprehensive set of tools to build conversational AI. It comprises of Bot Framework Composer, Bot Framework Solutions, Botkit, Bot Framework Emulator, Bot Framework Web Chat, Bot Framework Tools, Language Understanding, QnA Maker, Dispatch, Speech Services, Adaptive Cards, and Analytics. These are fairly large topics and beyond the scope of this book. In this chapter, we will focus on building a bot using Azure Bot Services.

Building Azure Service COVID-19 Bot[2]

1. Start with the Azure portal (portal.azure.com), and search for *bot services*, as shown in Figure 9-1.

Figure 9-1. *Azure portal – creating a bot services instance*

2. Select **Bot Services**, and you will see a list of different Bot Services offered by Microsoft and in the marketplace. Select **Web App Bot**, as shown in Figure 9-2.

---

[2]Due to COVID 19, Microsoft has created some open source templates as Rapid Bot deployment resources. These are for unemployment and other uses. The Microsoft Informational Bot templates are available at `https://microsoft.github.io/slg-covid-bot/`

Figure 9-2. *Azure portal – creating Web App Bot*

3.   Click **Create** to create a Web App Bot, which is going to use Azure
     Bot Services to help build, deploy, and manage the underlying
     activities. See Figure 9-3.

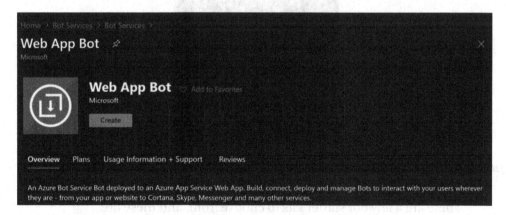

Figure 9-3. *Azure portal – creating Web App Bot*

4. Fill out the details for creating a bot, including the bot handle
    name, subscription, resource group, location, and pricing tier,
    and select a bot template. See Figure 9-4. Remember that you can
    always use your own existing App Service plan instead of creating
    a new one.

Figure 9-4. *Azure portal – creating Web App Bot*

There are a few bot starter kits to choose from, and these are
available in C# and Node.js. Echo Bot echoes the message, while
basic bot demonstrates the LUIS skills. See Figure 9-5. We also

have virtual assistant, LUIS Bot, and QnA Bot. In this case, since
we are focused on creating a COVID-19 bot, we will use the QnA
template.

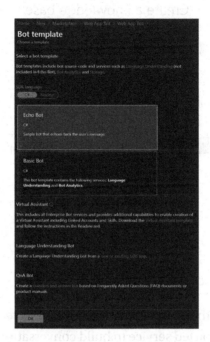

Figure 9-5. *Azure portal – bot templates*

5. In order to get the QnA Bot working, we need to create a Q&A
    knowledge base (KB). This knowledge base can be created here.
    Visit QnA Maker at `www.qnamaker.ai/Create`, and log in, using
    your Azure account to see the screenshot shown in Figure 9-6.
    Click **Create a QnA service**.

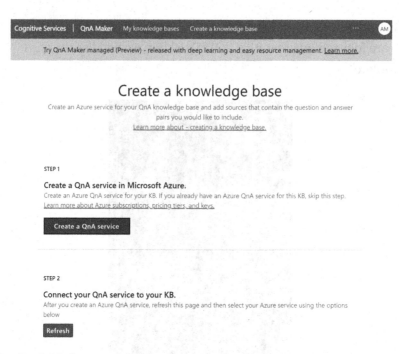

Figure 9-6. *QnA Maker – creating a knowledge base*

6. You will then see the Cognitive Services Create screen. QnA
    Maker is a cloud-based service to build conversational UI. Fill
    the following form (opens a new window of portal to create QnA
    Maker) with your subscription, resource group, and application
    service details. Click **Review + create** to continue.

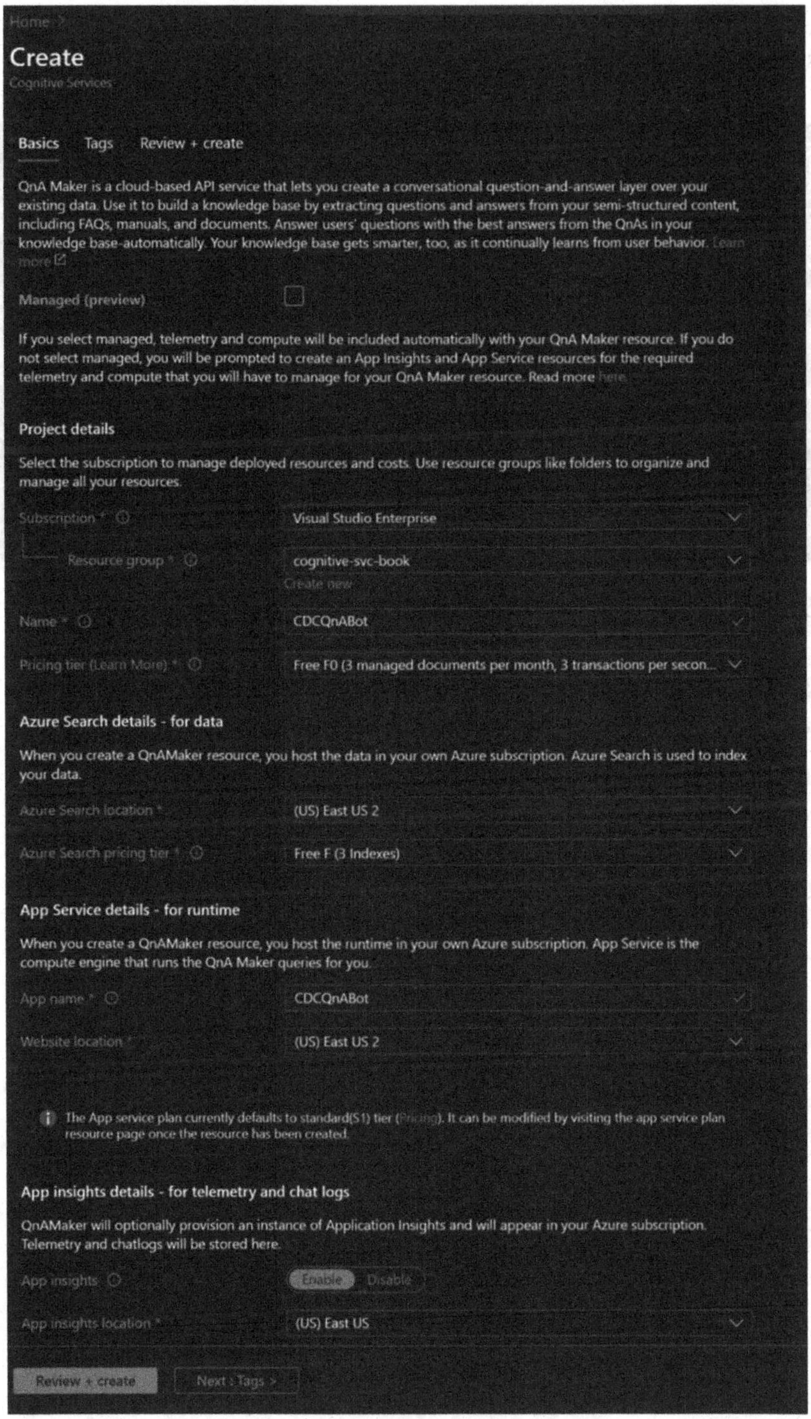

***Figure 9-7.** Azure portal – creating Cognitive Services*

7. Once the service is created and deployed, you will see the screen
    shown in Figure 9-8. Then proceed to the next level in QnA Maker.

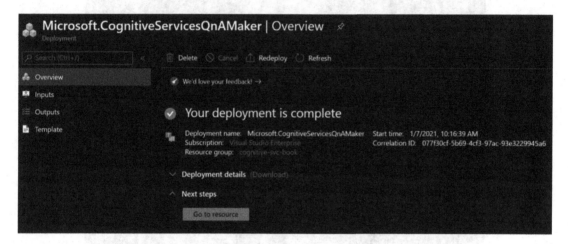

Figure 9-8. Azure portal – Cogntive Services deployment completion

8. Back at QnA Maker, step 2 is to connect the QnA service, which
    we just connected to our knowledge base. It usually takes around
    ten minutes after the resource is created for the runtime to be
    ready. Then can we select the service, and create the KB in QnA
    Maker portal. Click **Refresh** on the screen shown in Figure 9-9, to
    populate the respective fields.

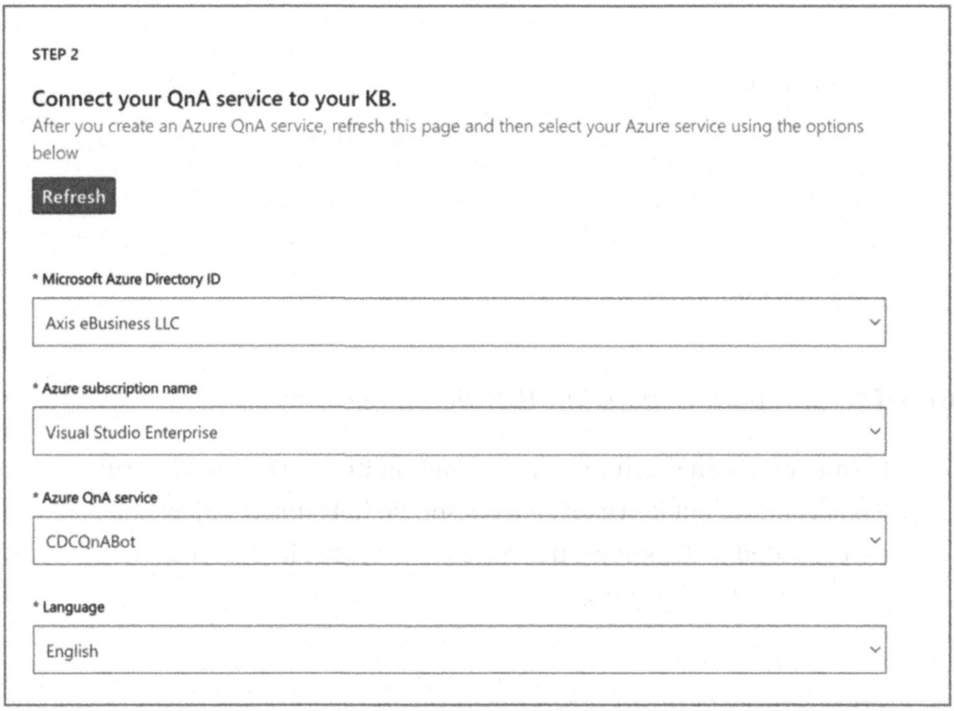

STEP 2

Connect your QnA service to your KB.

After you create an Azure QnA service, refresh this page and then select your Azure service using the options below

Refresh

* Microsoft Azure Directory ID

Axis eBusiness LLC

* Azure subscription name

Visual Studio Enterprise

* Azure QnA service

CDCQnABot

* Language

English

Figure 9-9. *QnA Maker – connect the QnA service to a knowledge base*

Once populated, you should name your knowledge base. We called it CDCQnABotKB. See Figure 9-10.

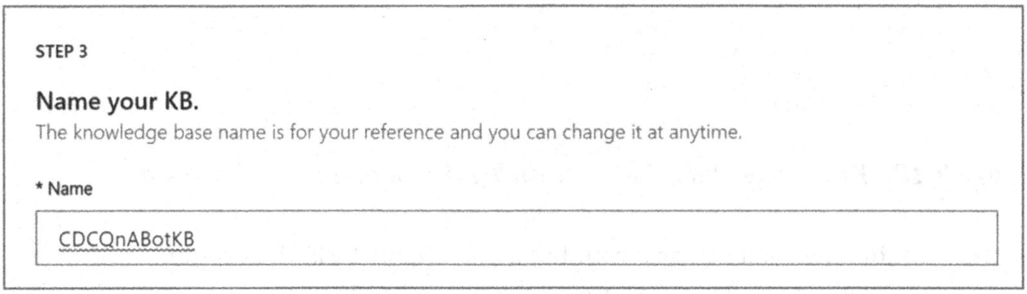

STEP 3

Name your KB.

The knowledge base name is for your reference and you can change it at anytime.

* Name

CDCQnABotKB

Figure 9-10. *QnA Portal – name the knowledge base*

And then finally, you would now make your KB (knowledge base), meaning that you will populate it with the contents. There are a few ways of uploading content to the knowledge base. First, you can provide a URL or files to read from. See Figure 9-11.

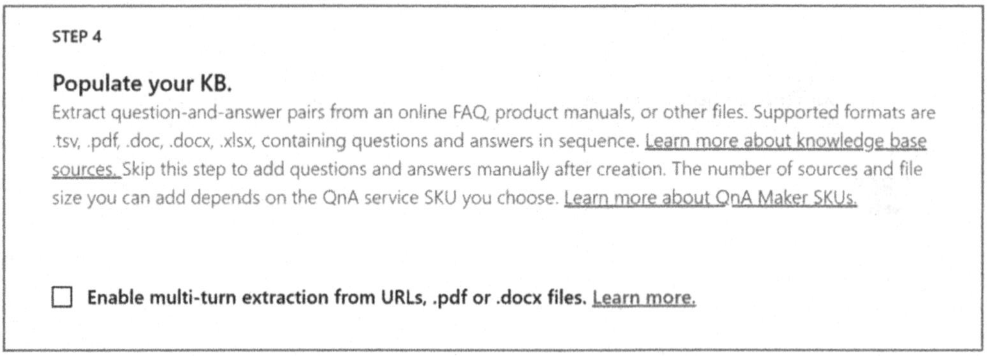

STEP 4
Populate your KB.
Extract question-and-answer pairs from an online FAQ, product manuals, or other files. Supported formats are .tsv, .pdf, .doc, .docx, .xlsx, containing questions and answers in sequence. Learn more about knowledge base sources. Skip this step to add questions and answers manually after creation. The number of sources and file size you can add depends on the QnA service SKU you choose. Learn more about QnA Maker SKUs.
Enable multi-turn extraction from URLs, .pdf or .docx files. Learn more.

Figure 9-11. *QnA portal - populate the knowledge base*

In this case, we are getting the information from CDC corona virus FAQs[3], and we put it in a text file (as shown in Figure 9-12), which we uploaded to the knowledge base via the Add File option. You can also use the PDF files.

```
CDC Q&A.txt
What is a novel coronavirus?

A novel coronavirus is a new coronavirus that has not been previously identified. The virus causing coronavirus disease 2019 (COVID-19), is not the same as the coronaviruses that commonly circulate among humans and cause mild illness, like the common cold.

Why is the disease being called coronavirus disease 2019, COVID-19?

On February 11, 2020 the World Health Organization announced an official name for the disease that is causing the 2019 novel coronavirus outbreak, first identified in Wuhan China. The new name of this disease is coronavirus disease 2019, abbreviated as COVID-19. In COVID-19, 'CO' stands for 'corona,' 'VI' for 'virus,' and 'D' for disease. Formerly, this disease was referred to as "2019 novel coronavirus" or "2019-nCoV".

There are many types of human coronaviruses including some that commonly cause mild upper-respiratory tract illnesses. COVID-19 is a new disease, caused by a novel (or new) coronavirus that has not previously been seen in humans. Top of Page

How does the virus spread?
```

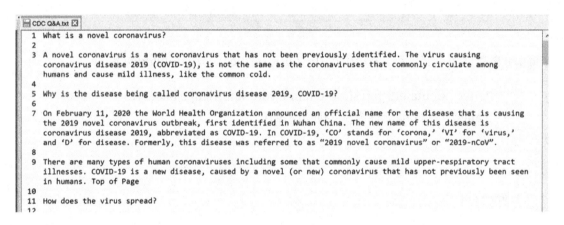

Figure 9-12. *Knowledge base file containing the cleaned up CDC data*

9. In this step, you can give your bot a personality by adding some predefined small talk questions with personas. The personas (shown in Figure 9-13) are provided as part of the chitchat capability in Azure Bot Services.

[3]www.cdc.gov/coronavirus/2019-ncov/faq.html

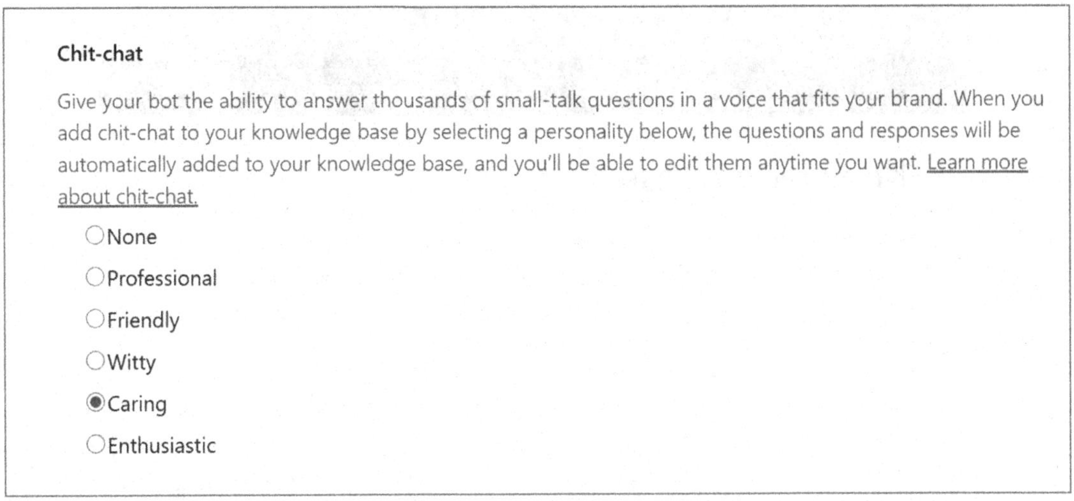

Chit-chat

Give your bot the ability to answer thousands of small-talk questions in a voice that fits your brand. When you add chit-chat to your knowledge base by selecting a personality below, the questions and responses will be automatically added to your knowledge base, and you'll be able to edit them anytime you want. Learn more about chit-chat.

○ None

○ Professional

○ Friendly

○ Witty

◉ Caring

○ Enthusiastic

Figure 9-13. *QnA portal – adding the chitchat module to the knowledge base*

10. The final step is to create your knowledge base. Click **Create your KB**, as shown in Figure 9-14.

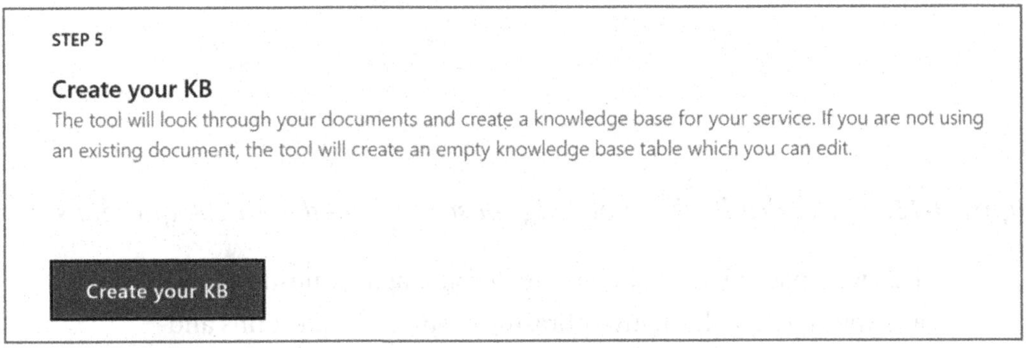

STEP 5

Create your KB

The tool will look through your documents and create a knowledge base for your service. If you are not using an existing document, the tool will create an empty knowledge base table which you can edit.

Create your KB

Figure 9-14. *QnA portal – create your knowledge base*

After a brief loading screen, you can now see all the questions and answers, as part of the QnA Maker knowledge base (as shown in Figure 9-15).

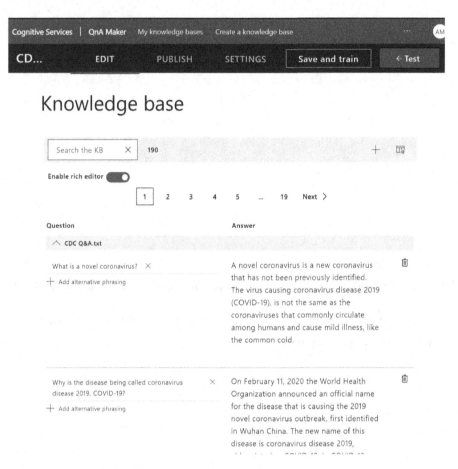

Figure 9-15. *QnA portal – the knowledge base populated with 190 questions*

This web-based IDE is quite useful for managing questions and answers, adding alternative phrasing, testing the questions and answers, and providing alternative answers (if needed). For example, in Figure 9-16, we searched for the keyword *swimming*, to see what the CDC guidelines are for swimming. In this case, it's just a keyword search, not a question and answer. See Figure 9-16.

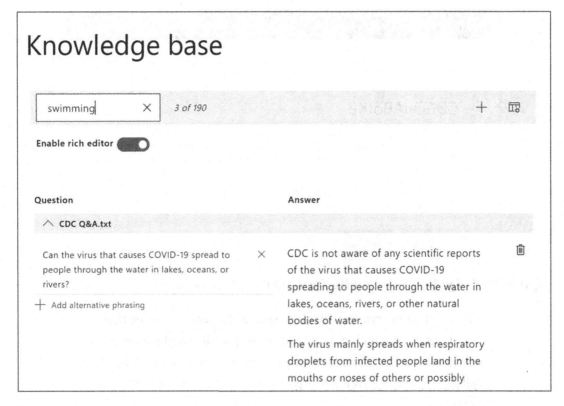

Figure 9-16. *QnA portal – searching the knowledge base*

11.   Here, you see the question and answer pair appear with the
      right answer. Notice that the question does not directly include
      the word *swimming*, but it implies the bodies of water, and the
      complete answer refers to the *swimming* key term. Now it's time
      to publish the serivce, for the world to see. Click the **Publish** tab
      in the top menu, and then click the **Publish** button (as shown in
      Figure 9-17).

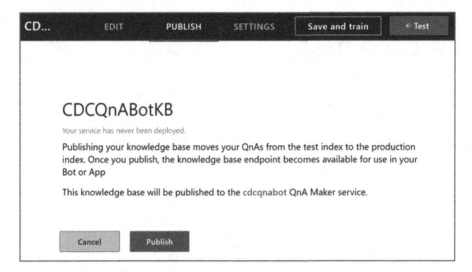

Figure 9-17. QnA portal – publishing the knowledge base

Providing that everything has gone smoothly, you would see the following screen which notifies that your question and answering knowledge base has been deployed as Azure Bot Services. It is available as a public API at this point. You can call it via postman or cURL.

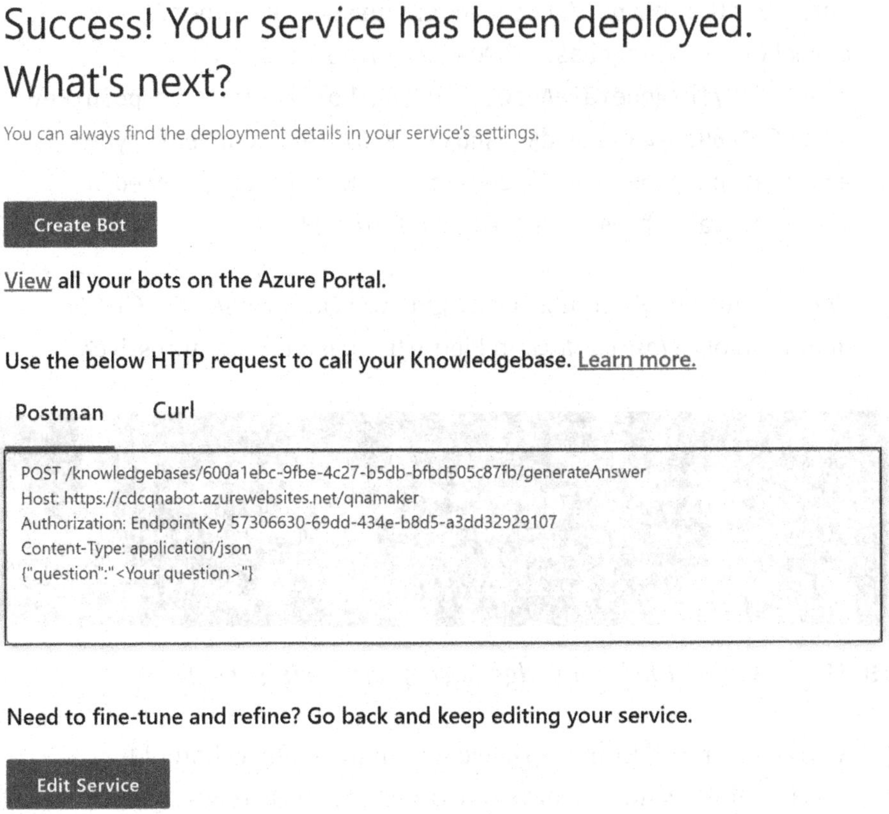

Success! Your service has been deployed. What's next?

You can always find the deployment details in your service's settings.

Create Bot

View all your bots on the Azure Portal.

Use the below HTTP request to call your Knowledgebase. Learn more.

Postman   Curl

```
POST /knowledgebases/600a1ebc-9fbe-4c27-b5db-bfbd505c87fb/generateAnswer
Host: https://cdcqnabot.azurewebsites.net/qnamaker
Authorization: EndpointKey 57306630-69dd-434e-b8d5-a3dd32929107
Content-Type: application/json
{"question":"<Your question>"}
```

Need to fine-tune and refine? Go back and keep editing your service.

Edit Service

Figure 9-18. *QnA portal – the knowledge base deployment screen*

Testing the KB from the QnA Maker portal is also a great way to understand how the final bot will respond.

12. In order to test the service out, we will ask the question, "Who is considered a close contact to someone with COVID-19?" We will need the endpoint and key, which will be available from the Azure Bot Services instance that we created earlier. Use the following code to make the cURL call:

```
curl -X POST https://cdcqnabot.azurewebsites.net/
qnamaker/knowledgebases/600a1ebc-9fbe-4c27-b5db-
bfbd505c87fb/generateAnswer -H "Authorization: EndpointKey
57306630-69dd-434e-b8d5-a3dd32929107" -H "Content-type:
application/json" -d "{'question':'Who is considered a
close contact to someone with COVID-19?'}"
```

The response JSON comes back right away, and we see the CDC's
definition of a *close contact*, in Figure 9-19. Isn't that impressive?

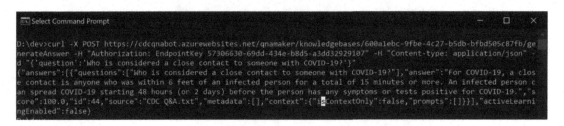

Figure 9-19. *Invoking the knowledge base query, using cURL*

13. As you may recall from step 4, we set out to create a chatbot for
     COVID-19. Now that we have our knowledge base ready, we can
     go back and create a bot by clicking the **Create Bot** button in the
     previous screen (see Figure 9-18) on the QnA portal. It will open
     up the Web App Bot screen, with a pre-populated QnA authkey.
     This means the bot is now connected to the QnA service.

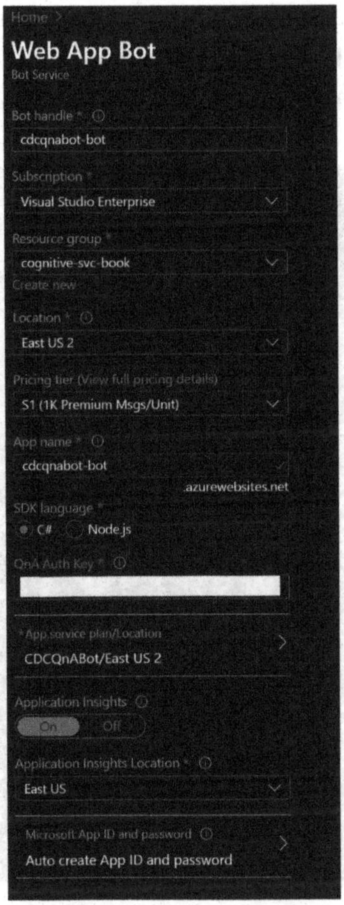

Figure 9-20. *Azure portal – creating the Web App Bot*

Verify the information, and then click **Create** to proceed. Your bot will be created, and once the deployment is successful, you will see the screen shown in Figure 9-21.

Figure 9-21. *Azure portal – Bot Services deployment successful*

14.   We now have the Web App Bot ready, along with the recipes to build, test, publish, connect with adapters, and evaluate the analytics as part of the one-stop Azure portal. See Figure 9-22.

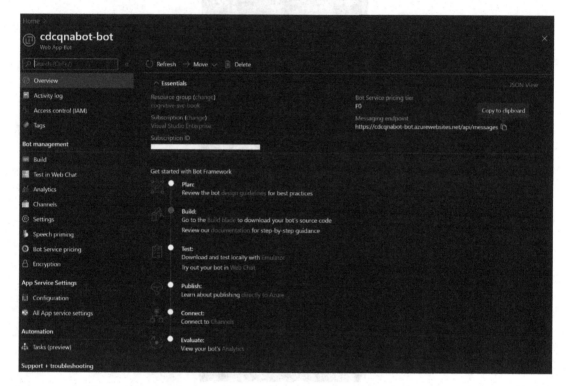

Figure 9-22.   *Azure portal – Web App Bot in Azure Bot Services*

You can also test the bot in the browser, by using the web chat feature. Click **Test in Web Chat** in the left pane, and you will see the screen to test the features. In this case, we will ask it questions about a funeral service, and you can see that without any prior training, it shows you the respective answers. See Figure 9-23.

Figure 9-23. *Azure portal – testing the Web Chat*

Before we move to the next section, let's make sure we are clear
on pricing. The Free tier allows 10K messages, but the SLA is not
guaranteed. With the Standard tier, there is a limited number
of messages, but with a 99.9% SLA guarantee. Check your
requirements to make sure you select the right pricing tier. See
Figure 9-24.

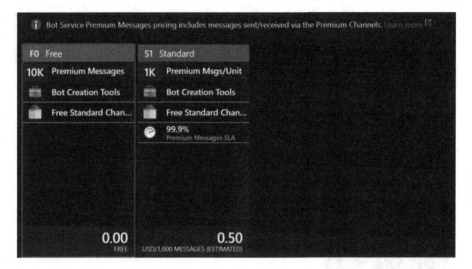

Figure 9-24. *Azure portal – pricing tier information*

In the earlier example, you saw how easy it is to build a QnA bot, by using QnA Maker and Azure Bot Services. But as you might have noticed, there is a lot going on in the background to make it happen. There is a lot of heavy lifting in terms of parsing natural language queries, retrieving data from text files, converting it into questions and answer pairs, being able to publish these queries as a service, and then answering them in real time. The good news is that all of this is hidden from us, and we don't have to build anything, thanks to Azure Bot Services.

In the next section, we will see how we can augment the bot knowledge base from the Web and how we can do interesting things with natural language questions.

Augmenting the Bot Knowledge Base from the Web

The standard CDC QnA is great. However, our fictitious client wants to make the COVID-19 bot more localized. Hillsborough county is the fourth most populous county in Florida, and they would like their FAQs[4] to be a part of this bot. Follow these steps:

1. Go to QnA Maker, and click the **My knowledge bases** link from the top bar. You will see your knowledge base CDCQnABotKB listed, as shown in Figure 9-25.

---

[4]Getting tested for COVID-19 www.hillsboroughcounty.org/en/residents/public-safety/emergency-management/stay-safe/getting-tested

My knowledge bases
Axis eBusiness LLC
All subscriptions
All services
Knowledge base name
Last modified
Last published
Sample Code
Azure service name
CDCQnABotKB
1/7/2021, 12:06:42 PM
1/7/2021, 12:05:44 PM
View Code
cdcqnabot

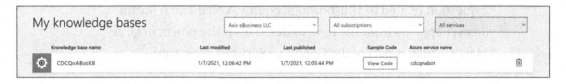

***Figure 9-25.** QnA portal – list of knowledge bases*

Click the knowledge base name (**CDCQnABotKB**) to proceed.

2. To add new resources, click **settings tab** in the QnA portal as part of your model. It will take you to the Manage knowledge base screen, as shown in Figure 9-26.

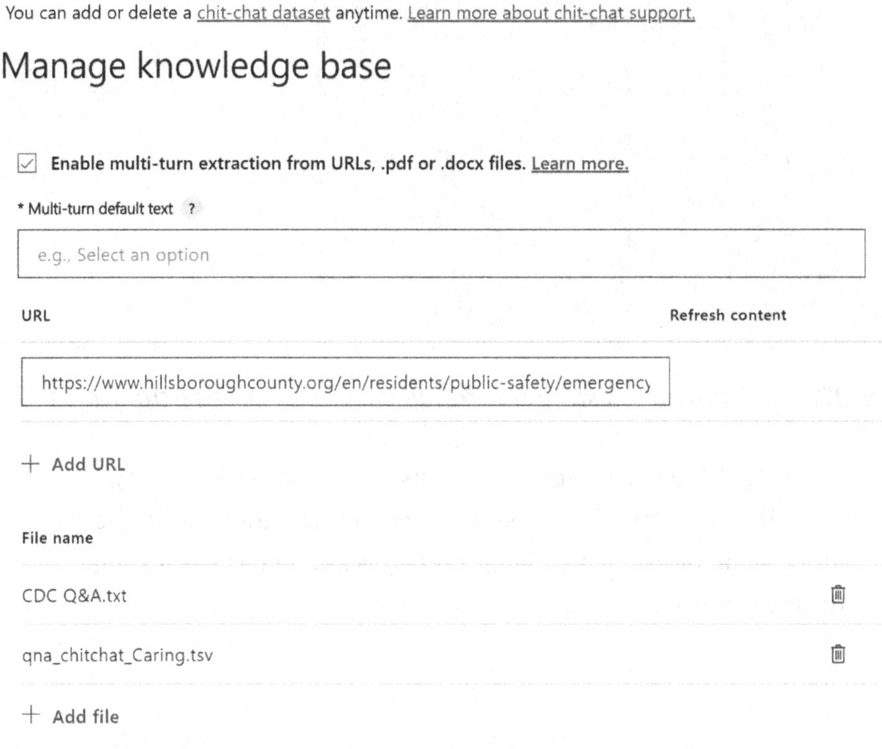
You can add or delete a chit-chat dataset anytime. Learn more about chit-chat support.
Manage knowledge base
Enable multi-turn extraction from URLs, .pdf or .docx files. Learn more.
* Multi-turn default text ?
e.g., Select an option
URL
Refresh content
https://www.hillsboroughcounty.org/en/residents/public-safety/emergenc
+ Add URL
File name
CDC Q&A.txt
qna_chitchat_Caring.tsv
+ Add file

***Figure 9-26.** QnA portal – manage knowledge base*

3. Add the Hillsborough county FAQs URL in the link (the URL is `www.hillsboroughcounty.org/en/residents/public-safety/emergency-management/stay-safe/getting-tested`). You can now click **Save and train** from the top menu. After that, click the blue **Test** button (as shown in Figure 9-27), and then ask

a question related to Hillsborough county. Voila! The relevant answer appears from the new source. This is the power of QnA Maker, which is part of the Cognitive Services ecosystem.

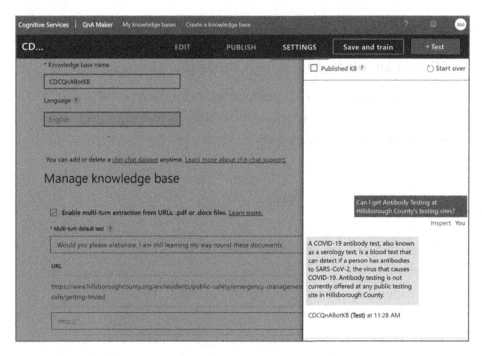

Figure 9-27. *QnA portal – testing the knowledge base for questions and answers*

However, if you closely look at the answer, it leaves a few things to be desired. First of all, it doesn't give you information about the address and hours of the Raymond James and Lee Davis testing locations. See Figure 9-28.

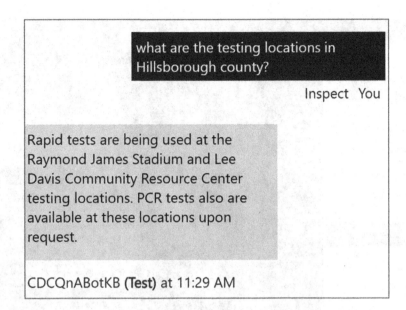

Figure 9-28. *QnA portal – testing the knowledge base for questions and answers*

4. We assume the problem is that this information is stored in the tables within the page, and it is not part of the extracted knowledge base. However, we can improve upon it. Click the **Inspect** link in Figure 9-28, and you will see the answers with their corresponding confidence scores (as shown in Figure 9-29).

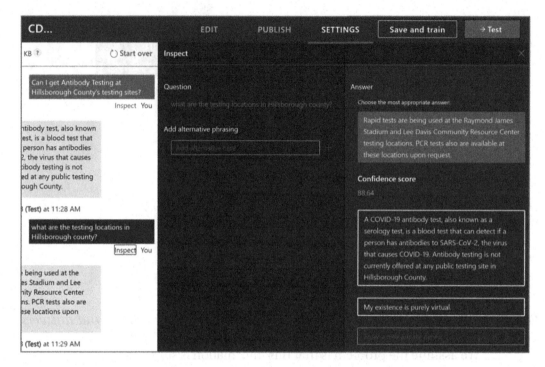

Figure 9-29. *QnA portal – editing the knowledge base for questions and answers*

This is an excellent place to test your questions, improve answers, add alternative phrasing for questions (other ways that these questions can be asked), and also add alternative answers. This is exactly what we did, as shown in Figure 9-30. We added the locations in the existing answers, and we also added some alternative phrasing.

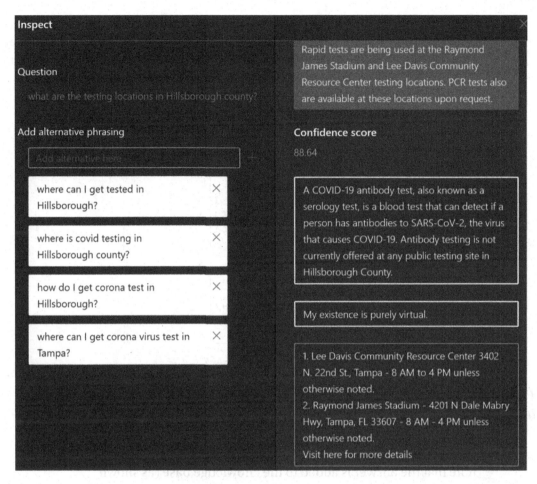

Figure 9-30. *QnA portal – editing the knowledge base for questions and answers*

5. Now you can click **Save and train** from the top menu, and then run the query again. See the answer in Figure 9-31.

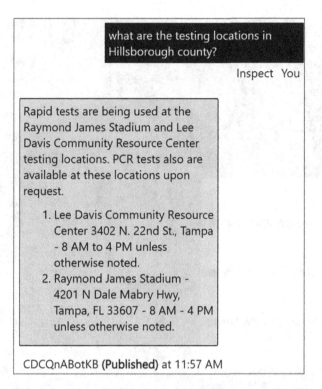

what are the testing locations in Hillsborough county?

Inspect You

Rapid tests are being used at the Raymond James Stadium and Lee Davis Community Resource Center testing locations. PCR tests also are available at these locations upon request.

1. Lee Davis Community Resource Center 3402 N. 22nd St., Tampa - 8 AM to 4 PM unless otherwise noted.
2. Raymond James Stadium - 4201 N Dale Mabry Hwy, Tampa, FL 33607 - 8 AM - 4 PM unless otherwise noted.

CDCQnABotKB (**Published**) at 11:57 AM

Figure 9-31. *QnA portal – knowledge base response with edited questions and answers*

Now the answer is detailed and exactly what we wanted. You will note that the answer is added to the knowledge base (as shown in Figure 9-32), with the source identified as *editorial*. This helps identify the source citation of the answers.

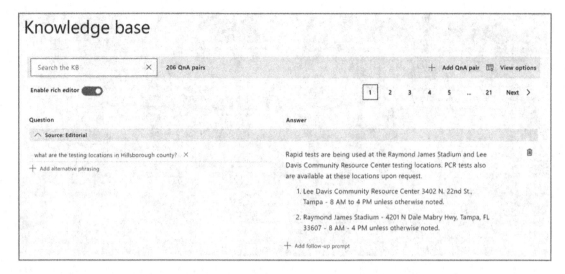

Figure 9-32. *QnA portal – knowledge base updated with editoral answers*

6. Now that the bot knowledge base is trained and running, we can call the API. Unlike the earlier cURL instance, this time we would set the debug to be true, as shown in the following code:

```
curl -X POST https://cdcqnabot.azurewebsites.net/qnamaker/
knowledgebases/600a1ebc-9fbe-4c27-b5db-bfbd505c87fb/
generateAnswer -H "Authorization: EndpointKey 57306630-69dd-
434e-b8d5-a3dd32929107" -H "Content-type: application/
json" -d "{'question':'can I go to a funeral safely?',
'Debug':{'Enable':true}}"
```

This enables us to get a larger question answer set, along with the debug flag items, such as the associated ranks, synonyms, and probabilities of answers, which help us understand the responses. See the cURL response in Figure 9-33.

Figure 9-33. The cURL response with debug from the Bot Services API

7. Lastly, the analytics capabilities of Azure Bot Services are phenomenal. You can trace the number of users, activities, the channels (Alexa, Cortana, and so on), and check out the retention of users. See Figure 9-34. This screen is available on Azure portal on the bot resource, conversational analytics tab, or blade.

● Alexa   ● Cortana   ● Direct Line   ● Direct Line Speech   ...   ∨   12/7/2020 - 1/7/2021   ∨

GRAND TOTALS

2   **32**

Users   Activities

RETENTION - % USERS WHO MESSAGED AGAIN (LAST 10 DAYS)

Date	Users				Days later					
		1	2	3	4	5	6	7	8	9	10
12/27/2020	0	0%	0%	0%	0%	0%	0%	0%	0%	0%	0%
12/28/2020	0	0%	0%	0%	0%	0%	0%	0%	0%	0%	
12/29/2020	0	0%	0%	0%	0%	0%	0%	0%	0%		
12/30/2020	0	0%	0%	0%	0%	0%	0%	0%			
12/31/2020	0	0%	0%	0%	0%	0%	0%				
1/1/2021	0	0%	0%	0%	0%	0%					
1/2/2021	0	0%	0%	0%	0%						
1/3/2021	0	0%	0%	0%							
1/4/2021	0	0%	0%								
1/5/2021	0	0%									

Figure 9-34. Azure portal – analytics about user engagement with the chatbot

Conclusion and Summary

Digital conversations are on the rise, and bots are our first line of defense in this barrage of queries, which require instant gratification. In this chapter, we scratched the surface of what Azure Bot Services and Bot Framework are capable of. We started by explaining the key aspects of Azure Bot Services, we created a COVID-19 bot using Azure Bot Services, and then we tested it using Azure Bot Builder SDK and QnA Maker.

Now that you built a fundamental understanding of Azure Bot Services, you can start exploring and use these technologies for your own business use cases. Natural language conversations have tons of implementations. From healthcare to finance, retail, and hospitality, the industries are brimming with use cases to provide improved customer experiences for your users.

Keep chatting!

References and Further Reading

Build bots with Bot Framework

https://dev.botframework.com/

Bot Framework Docs

https://docs.microsoft.com/bot-framework/

Bot Builder SDK

https://github.com/Microsoft/BotBuilder

Asked and Answered: Building a Chatbot to Address Covid-19-Related Concerns

https://catalyst.nejm.org/doi/full/10.1056/cat.20.0230

Quickstart: Create, train, and publish your QnA Maker knowledge base

https://docs.microsoft.com/azure/cognitive-services/qnamaker/quickstarts/create-publish-knowledge-base

Bot Builder Samples

https://github.com/Microsoft/BotBuilder-Samples/blob/main/README.md

Introducing QnA Maker managed: now in public preview

https://techcommunity.microsoft.com/t5/azure-ai/introducing-qna-maker-managed-now-in-public-preview/ba-p/1845575

CHAPTER 10

Azure Machine Learning

Q: Why did Ed get attacked in the haunted jungle?

A: Because other people are sensible enough to not enter random forests.

That's a machine learning joke at the expense of our co-author, but it makes an excellent segue into customized algorithms such as random forests. (A random decision forest classifies tasks in decision trees to output mean or average predictions/ regressions. Ed's revenge is that he explains jokes.) So far, in this book, you have done an extensive review of Cognitive Services, which provide a set of prebuilt algorithms, models, and recipes in a managed and well-organized manner. These services enable and accelerate machine learning development and make data scientists deliver products and services faster and more effectively without reinventing the wheel.

However, real-world business use cases exist in data science and machine learning, beyond using prebuilt APIs. You have to be able to build your own models, try out new datasets, augment existing models, and create your own algorithms. Rest assured Microsoft Azure has extensive capabilities to address each one of these customized needs. Azure Machine Learning is an integral part of the Microsoft AI ecosystem, which helps machine learning engineers, AI enthusiasts and professionals, and citizen data scientists alike, to implement intelligent solutions. The end-to-end machine learning life cycle comprises of not only building, training, and deploying machine learning models, but it's also maintaining them in the long run. You must monitor the model drift and decay, scale based on demand, and complete challenger champion tests to ensure that the best model wins. The Azure Machine Learning platform helps bring together a diverse set of these services, to support the full machine learning life cycle.

© Ed Price, Adnan Masood, and Gaurav Aroraa 2021
E. Price et al., *Hands-on Azure Cognitive Services*, https://doi.org/10.1007/978-1-4842-7249-7_10

This chapter will help you understand Azure Machine Learning ecosystem offerings and how to train an application to learn without being explicitly programmed, which is the goal of machine learning. So far, you have mostly seen Azure Cognitive Services in action, but in this chapter, we will review the entire Azure Machine Learning ecosystem to provide you a big picture of life beyond Cognitive Services. You will learn

- The Azure Machine Learning stack and offerings

- Building models with no code, using the Azure Machine Learning designer

- Building ML models in Python and deploying them using Azure notebooks

- Deploying and testing models, with Azure Machine Learning

Let's dive in.

Azure Machine Learning Stack

Azure Machine Learning stack is a cloud platform hyperscaler force to reckon with. It contains the entire end-to-end machine learning platform capabilities, from data ingestion to model deployment and management. For instance, to bring in the datasets for running predictive analytics, and to include data science experiments, Azure offers storage repositories, like Azure Blob Storage and Azure Data Lakes Storage. For the machine learning compute resources, there are options of individual virtual machines (VMs), Spark clusters with HD Insight, or Azure Databricks.

Figure 10-1 outlines the different offerings within the Azure Machine Learning stack and what should be used when. The focus of this chapter is Azure Machine Learning.

Cloud options	What it is	What you can do with it
Azure Machine Learning	Managed platform for machine learning	Use a pretrained model. Or, train, deploy, and manage models on Azure using Python and CLI
Azure Cognitive Services	Pre-built AI capabilities implemented through REST APIs and SDKs	Build intelligent applications quickly using standard programming languages. Doesn't require machine learning and data science expertise
Azure SQL Managed Instance Machine Learning Services	In-database machine learning for SQL	Train and deploy models inside Azure SQL Managed Instance
Machine learning in Azure Synapse Analytics	Analytics service with machine learning	Train and deploy models inside Azure SQL Managed Instance
Machine learning and AI with ONNX in Azure SQL Edge	Machine learning in SQL on IoT	Train and deploy models inside Azure SQL Edge
Azure Databricks	Apache Spark-based analytics platform	Build and deploy models and data workflows using integrations with open-source machine learning libraries and the MLFlow platform.

Figure 10-1. *Azure Machine Learning key offerings – from* https://docs. microsoft.com/ azure/architecture/data-guide/technology-choices/data-science-and-machine-learning

Enterprise readiness and governance requires data protection, which is offered by Azure Key Vault, to manage and secure credentials. Repeatable and reproducible experimentation helps achieve consistent results and require a managed and properly versioned set of models and dependencies. With Microsoft Azure, you can use Docker containers, and with ACI and AKS, you can manage these repositories to build and deploy as needed. In the context of machine learning, the salient capabilities are shown in Figure 10-2.

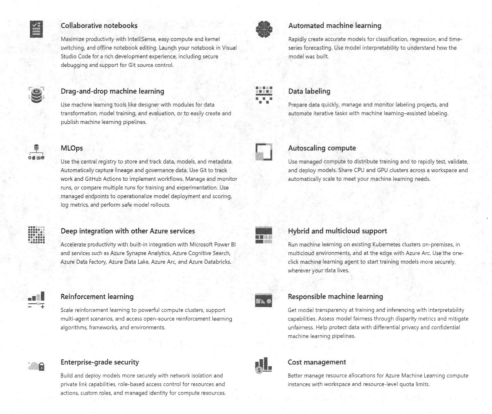

Figure 10-2. *Azure Machine Learning key service capabilities – from* `https:// azure.microsoft.com/services/machine-learning`

The following provides a brief overview of what these entail, which you will see as part of Azure Machine Learning studio, the machine learning one-stop shop for Azure data scientists. The highest item in the hierarchy of Azure Machine Learning is a workspace, as shown in Figure 10-3. It contains a variety of different elements, such as experiments, pipelines, datasets, models, and so on, as we will discuss in the following.

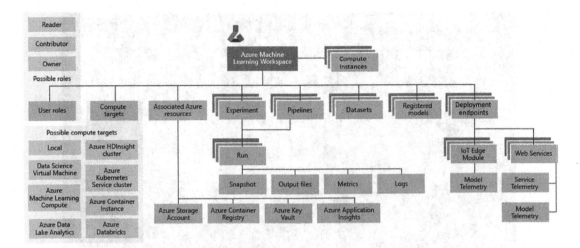

Figure 10-3. *Azure Machine Learning workspace – from* `https://docs.` `microsoft.com/azure/machine-learning/concept-workspace`

1. **Endpoints** – A machine learning model needs to be consumed, and there are few ways to do that. One of the popular ways to deploy a machine learning model is via a web service, which contains one or more endpoints. An endpoint is an address where a model can be reached and invoked via a well-defined protocol, such as RESTful APIs over HTTP. (*REST* stands for representational state transfer.)

2. ***Azure notebooks*** – Azure notebooks provide an integrated working environment for data scientists to access files, folders, storage, and compute, all within the Azure workspace. We have seen Jupyter notebooks in Chapter 6, and Azure notebooks are similar; they just use Azure compute and storage, which we will explore shortly. Azure notebooks provide an excellent place to collaborate and work with datasets, models, and experiment in one place. See Notebooks in Figure 10-4.

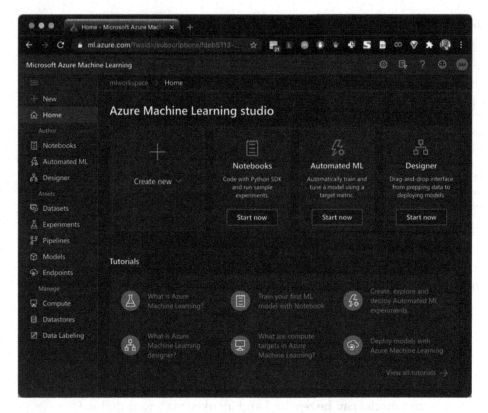

Figure 10-4. *Azure Machine Learning studio – key offerings*

3. **Automated Machine Learning** – Automated machine learning (or
    AutoML) is the capability to apply machine learning techniques,
    to find out the best models for a problem. This includes
    automated feature engineering, neural architecture search, and
    hyperparameter optimization. Azure AutoML takes the guess work
    out of finding out the best hyperparameters, and it can help you
    find the best model to use, with a no-code approach.

4. **Datasets** – The management of machine learning assets, such
    as models and datasets, can be a challenging problem for ML
    practitioners. With Azure datasets, you can easily share and track
    data across multiple machine learning workflows, while a single
    copy of data stays in storage. The data is accessible without the
    concerns of connectivity and connection strings, for example.

5. **Experiments** – Experimentation and repeatability is what puts science in *data science*, to be able to create and experiment with datasets and algorithms. The Experiments section of Azure Machine Learning keeps track of your machine learning experiments and runs so you can replay the pipelines.

6. **Pipelines**[1] – *Pipeline* is possibly the most overloaded term in machine learning, which typically means data extraction, model development and training, tuning, production release, and then monitoring and review, or some combination of these activities. In the context of Azure Machine Learning, it is the executable workflow of these machine learning tasks that is repeatable, and that can perform data operations (such as normalization, transformation, featurization, validation, and so on), model building and training (such as hyperparameter tuning, automatic model selection, model testing, and validation), and model deployment (such as deployment and batch scoring). Model pipelines offer great features, like the following:

   a. Unattended execution

   b. Diverse computational resource accessibility

   c. Reproducibility and reusability

   d. Tracking and versioning

7. **Models** – Models are the bread and butter of machine learning: they encapsulate the learning from the data, by using the prescribed algorithm. In the Azure Machine Learning workspace, you can register, create, manage, and track models. Any model you built in other parts of the workspace also gets registered in this repository and can be used across the platform.

8. **Compute** – A machine learning model, an automated machine learning process, and a Jupyter notebook, all need a place to run (to perform a computation). This is exactly what Azure Compute

---

[1] Azure Machine Learning Pipeline: `https://github.com/Azure/MachineLearningNotebooks/tree/master/how-to-use-azureml/machine-learning-pipelines`

provides – here you can create different types of compute targets, as seen in Figure 10-5. This table elaborates on different training targets for workloads. For machine learning workloads, you can use local computers, a compute cluster, remote virtual machines, and a variety of other training targets.

Compute target	Used for	GPU support	FPGA support	Description
Local web service	Testing/debugging			Use for limited testing and troubleshooting. Hardware acceleration depends on use of libraries in the local system.
Azure Machine Learning compute instance web service	Testing/debugging			Use for limited testing and troubleshooting.
Azure Kubernetes Service (AKS)	Real-time inference	Yes (web service deployment)	Yes	Use for high-scale production deployments. Provides fast response time and autoscaling of the deployed service. Cluster autoscaling isn't supported through the Azure Machine Learning SDK. To change the nodes in the AKS cluster, use the UI for your AKS cluster in the Azure portal. AKS is the only option available for the designer.
Azure Container Instances	Testing or development			Use for low-scale CPU-based workloads that require less than 48 GB of RAM.
Azure Machine Learning compute clusters	Batch inference	Yes (machine learning pipeline)		Run batch scoring on serverless compute. Supports normal and low-priority VMs.
Azure Functions	(Preview) Real-time inference			
Azure IoT Edge	(Preview) IoT module			Deploy and serve ML models on IoT devices.
Azure Data Box Edge	Via IoT Edge		Yes	Deploy and serve ML models on IoT devices.

Figure 10-5. *Azure Machine Learning compute targets*

9. **Data Store** – Data stores are the repositories for data; they are the containers for datasets, such as Azure Storage, Azure Data Lake Storage, Azure SQL Database, and Azure Databricks File System (DBFS). For system storage, Azure Machine Learning uses an Azure blob container and file container.

10. **Labeling** – Labeling and annotation are important parts of the data science life cycle. Using the Azure Machine Learning labeling features, you can label datasets by creating labeling projects, managing workflows, and farming it out to multiple labelers, and you can also use machine learning to help you out.

You will learn more about these components, shortly, as you see them in action. As we move toward our journey to build a machine learning model, it is important to know the right algorithm to use for building the models.

Along with some well-articulated, concise, and managed documentation, the Azure Machine Learning team also offers a highly useful cheat sheet (aka.ms/mlcheatsheet), which assists in determining the best machine learning algorithm. The cheat sheet is shown in Figure 10-6.

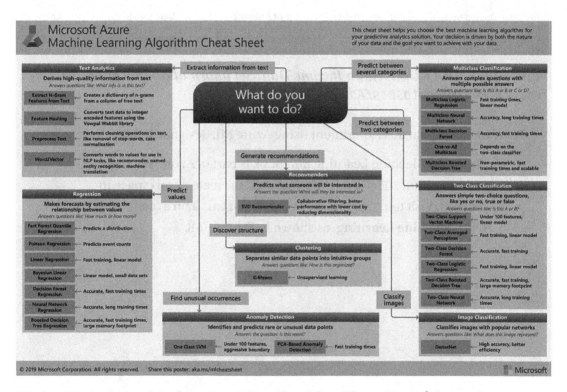

Figure 10-6. *Azure Machine Learning Algorithm Cheat Sheet (aka.ms/mlcheatsheet)*

Whether you want to predict between categories, discover a structure, or extract information from the text, in the following flow, you can find out the best algorithm, based on your needs.

Hello World with Azure Machine Learning

In the book so far, you have mainly worked with prebuilt libraries and APIs. In this next example, we will show how you can take the UCI Census Adult Income Data Set[2] and build a Jupyter notebook to predict if the income exceeds $50K per year, based on the census data. See Figure 10-7 for information about this dataset.

Data Set Characteristics:	Multivariate	Number of Instances:	48842	Area:	Social
Attribute Characteristics:	Categorical, Integer	Number of Attributes:	14	Date Donated	1996-05-01
Associated Tasks:	Classification	Missing Values?	Yes	Number of Web Hits:	2038956

Figure 10-7. *UCI Census Adult Income Data Set information (`https://archive.ics.uci.edu/ml/datasets/Adult`)*

Then we will repeat this experiment using Azure ML designer. Follow these steps:

1. The Azure Machine Learning notebook is part of Azure Machine Learning studio. However, first you need to create a workspace to proceed. Visit the Azure portal (`http://portal.azure.com`), and select **Machine Learning**, as shown in Figure 10-8.

---

[2]`http://archive.ics.uci.edu/ml/datasets/Adult`

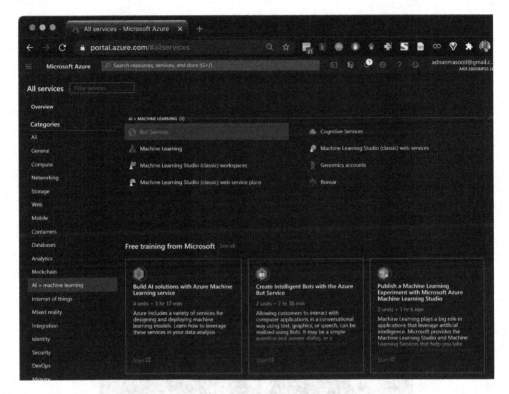

Figure 10-8. *Azure portal – creating a machine learning workspace*

2. Fill out the details about the subscription, workspace name,
    region, and storage account. You can leave most of these as the
    default settings, unless you have specific use cases in networking
    or containers you would like to address. Once completed, click
    Review + create to proceed (as shown in Figure 10-9).

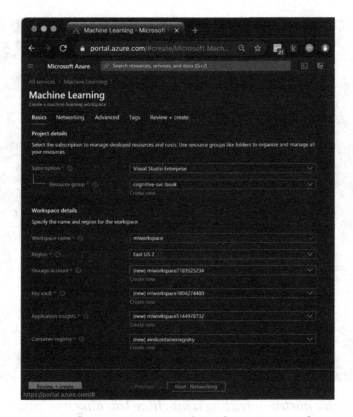

Figure 10-9. *Azure portal – creating a machine learning workspace*

3. Once the deployment completes, you will see the message shown in Figure 10-10. To continue, click **Go to resource**.

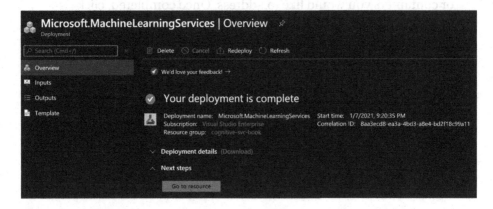

Figure 10-10. *Azure portal – creating a machine learning workspace*

4.   In the resource screen, you will see the summary of the created
     workspace. Now you are ready to proceed with Machine Learning
     studio. You can either click the Studio web URL or click the
     Launch studio button, which will take you to the Azure Machine
     Learning studio.

***Figure 10-11.** Azure portal – creating a machine learning workspace*

Azure Machine Learning studio   You might notice that there is another flavor to
the Machine Learning studio, called Azure Machine Learning studio (classic)[3]. This
is not to be confused with the current, modernized Azure Machine Learning studio
interface, which includes a designer, automated machine learning, and Jupyter
notebook capabilities. The Azure Machine Learning studio (classic) was the original
drag-and-drop UI for building machine learning models, created in 2015. Most of
its features are available as part of the designer, but there is a whole lot more in
the modern Azure Machine Learning studio environment.

---

[3]https://studio.azureml.net/

5. This is the main screen for Azure Machine Learning studio, which can now be accessed via the ML URL (ml.azure.com). It comprises of different capabilities and skills that you read about earlier. Let's click the notebooks' **Start now** button, to create a simple *hello world* notebook.

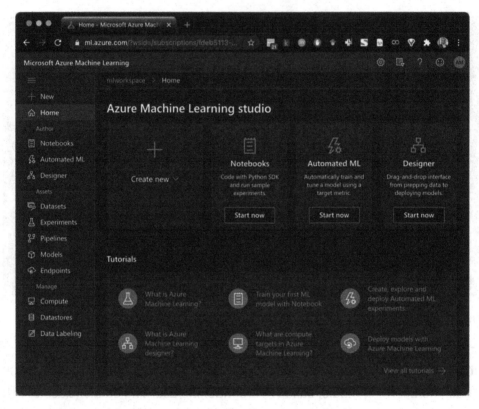

Figure 10-12. *Azure Machine Learning studio homepage*

You have an option to explore sample notebooks or to create a new Jupyter notebook file. As a baby step, let's create a new file, as shown in Figure 10-13. This will make it easy for you to understand the underlying dependencies.

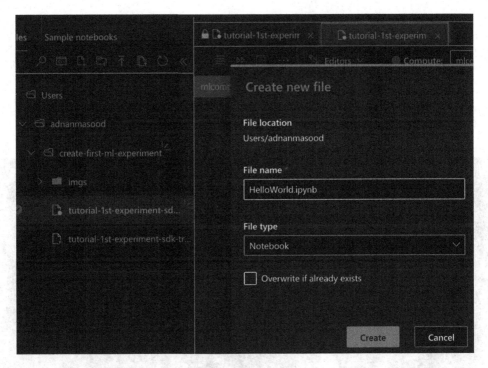

Figure 10-13. *Azure Machine Learning – creating a Jupyter notebook*

6.   Once you create a new Jupyter notebook, which is empty by default, you will notice the "No computes found" message, on the top right (as shown in Figure 10-14). If you try to run this notebook, you will get an error message stating that your notebook is currently not connected to a compute. It requires you to switch to a running compute or to create a new compute to run a cell. This simply means that you need a compute resource to run a notebook. Click the **Create compute** button to create a resource for the notebook.

Figure 10-14. *Azure Machine Learning – creating compute*

7. The compute instance screen helps you create a compute
    instance, in this case a virtual machine (VM). You can choose
    the VM type. If you have a deep learning–focused workload,
    use a GPU. Otherwise, a CPU should serve you fine. There are
    some prebuilt VMs, with their costs listed next to them – select a
    resource, as needed (see Figure 10-15).

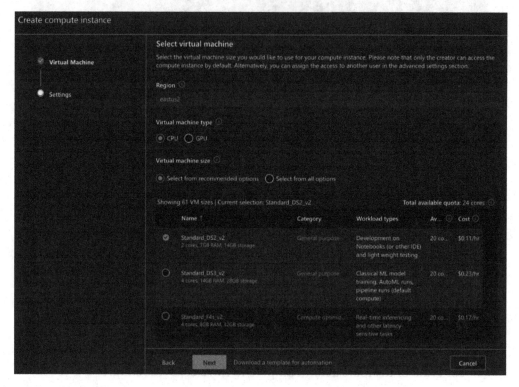

Figure 10-15. *Azure Machine Learning studio – selecting a VM*

For this simple example, we would choose a DS2 v2 machine with
the following specs (as shown in Figure 10-16). This is overkill
for the task at hand, printing *Hello World*, but we can reuse this
resource for our next task of classification.

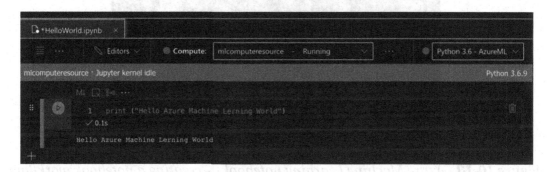

Figure 10-16. Azure Machine Learning studio – configuring the VM settings

8. Once you have chosen the VM and created the compute name,
    now it's time to run the notebook. Voila! See Figure 10-17.

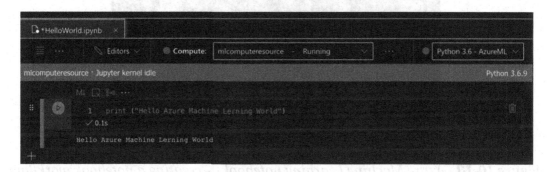

Figure 10-17. Azure Machine Learning studio – Hello World notebook

This example seems trivial, but it has prepared you with all the prerequisites for what is coming next, an income classification model that you build, by using Azure Machine Learning service and Jupyter notebook. Now, with all the dependencies out of the way, let's look into the census income classification problem.

Creating a Machine Learning Model for Classification

The UCI Census Adult Income Data Set[4] has been around for a while. It is used for a variety of different use cases, including to predict if the income exceeds $50K per year, based on the census data.

1. Create a notebook by clicking the **Create** button, as shown in Figure 10-18.

Figure 10-18.   *Azure Machine Learning notebooks – creating a notebook workflow*

Name the notebook AdultCensusClassifer.ipynb. In the File type drop-down box, select **Notebook** (as shown in Figure 10-19).

---

[4]http://archive.ics.uci.edu/ml/datasets/Adult

Figure 10-19. *Azure Machine Learning notebooks – creating a notebook workflow*

2. Upload the *adult.csv* and *adult_test.csv* datasets, which are part of the book's code repository. These files contain the adult dataset in the CSV format, as shown in Figure 10-20.

Figure 10-20. *Azure Machine Learning – uploading the dataset*

You will need to verify that you trust the contents of these files. Click the **I trust contents of these files** checkbox to give your consent, as shown in Figure 10-21. Any unsecure uploads can cause security and governance issues to your repositories.

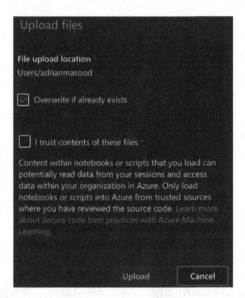

Figure 10-21. *Azure Machine Learning notebooks – uploading the dataset*

You can click the uploaded dataset to view the CSV file in the browser, as shown in Figure 10-22.

Figure 10-22. *Azure Machine Learning notebooks – viewing the dataset*

3. The experiment will try to apply different classification algorithms to the dataset, and then it will generate the accuracy scores. The code segment, as shown in Figure 10-23, loads the adult dataset, the adult test dataset, and the prerequisites and dependencies.

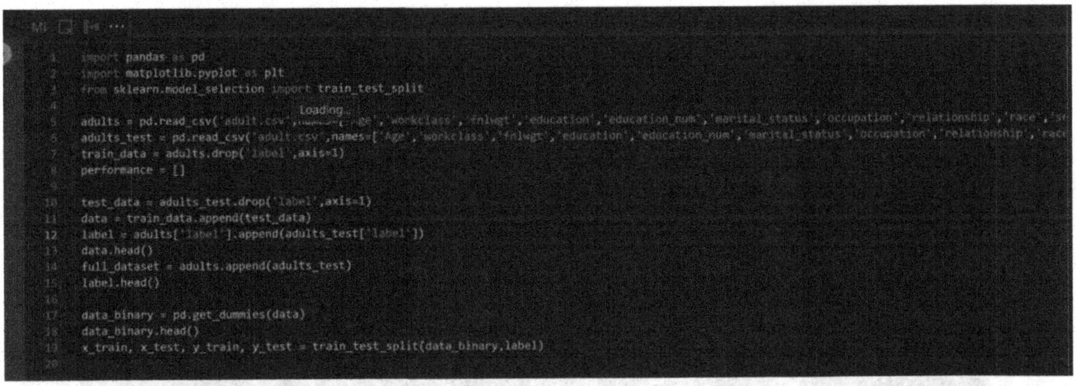

Figure 10-23. *Azure Machine Learning notebooks – code for running prerequisites and loading the datasets*

4. Let's run the logistic regression classifier on the dataset and then calculate the scores. The sklearn library makes the interface very consistent and easy to work with. Toward the end, we print the accuracy scores and append it to the performance array (as shown in Figure 10-24).

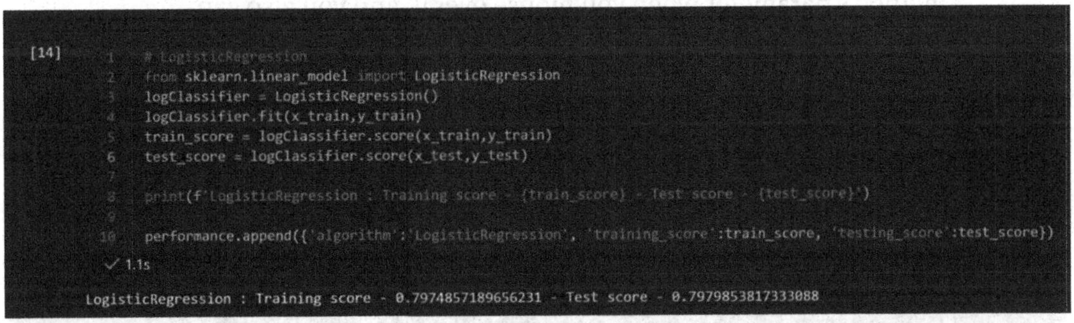

Figure 10-24. *Azure Machine Learning notebooks – code for logistic regression*

5. We repeat the same experiment with Gaussian Naïve Bayes, K neighbors, random forests, and support vector machines. In Figure 10-25, you can see that random forest has the best results for both training and test data.

```
[7]:  #Performance

      performance_df = pd.DataFrame(performance)
      performance_df
```

[7]:		algorithm	training_score	testing_score
	0	Gaussian Naive Bayes	0.795991	0.792887
	1	LogisticRegression	0.798305	0.795160
	2	K Neighbors	0.864724	0.813095
	3	Random Forests	0.999980	0.960138
	4	Support Vector Machine	0.854958	0.846692

Figure 10-25. *Azure Machine Learning notebooks – results for experiments*

Before we conclude, the notebook's IntelliSense, a code-completion aid, is a wonderful feature that gives you the look and feel of an IDE in the browser. In Figure 10-26, you can see the method's parameter when you mouse over it, and you also can use autocomplete.

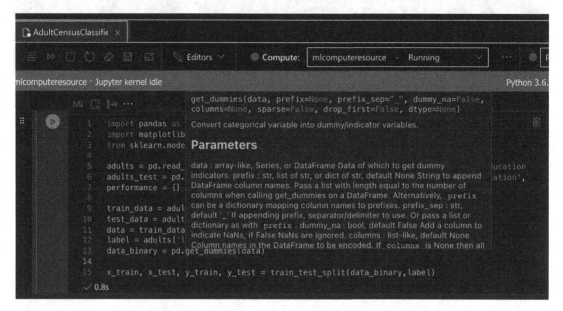

Figure 10-26. *Azure Machine Learning notebooks – IntelliSense and autocomplete*

This experiment provided you insight into how you can import a dataset into Azure notebooks and run multiple algorithms.

Azure Notebooks had a preview version (`https://notebooks.azure.com/`), which is now retired, and these notebooks are not an integrated part of Azure Machine Learning service. If you are looking for a free alternative for research and development purposes, Google Colab (`https://colab.research.google.com/notebooks/intro.ipynb`) provides an excellent choice. However, the service is *stateless*, which means upon every session expiration, you would need to reload the datasets.

Classification Model Using Azure Machine Learning Designer

In the earlier example, we built a classification model, by using Azure Machine Learning notebooks. This was a code-heavy approach that required us to write Python code for different algorithms. Now, we will follow a low-code approach, by using the Azure Machine Learning designer, as shown in Figure 10-27.

In the left pane of the Azure Machine Learning workspace portal, click Designer. The designer homepage opens, where you can see the collection of pipelines (shown in Figure 10-27).

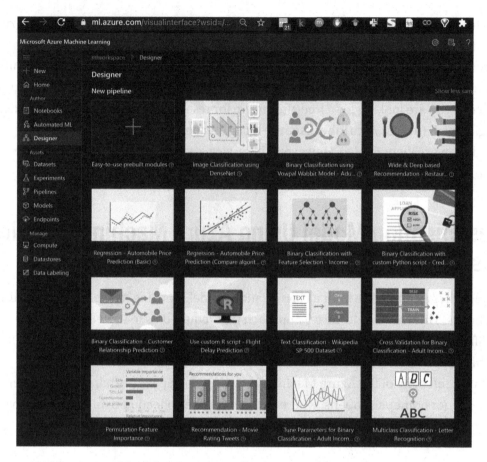

Figure 10-27. *Azure Machine Learning designer – main screen*

The Microsoft Azure team provides a set of basic and advanced recipes to learn how to use the designer. The example pipelines and datasets for Azure Machine Learning designer include regression examples, like automobile price prediction, classification use cases (such as binary classification with feature selection for income), credit risk, flight delay, customer relationship prediction, text classification with Wikipedia SP 500 Dataset, and computer vision–based Image Classification, using DenseNet. The examples also include recommendation engines, such as a restaurant rating prediction, movie ratings based on tweets, and binary classification using Vowpal Wabbit Model for Adult Income Prediction.

To build a classification model by using Azure Machine Learning designer, follow these steps:

1. In this example, we will use the Adult Census Income dataset. Open up the example by double-clicking it, from the home screen, or build it from scratch. To continue, click the **Designer** tab (in the left pane). See Figure 10-28.

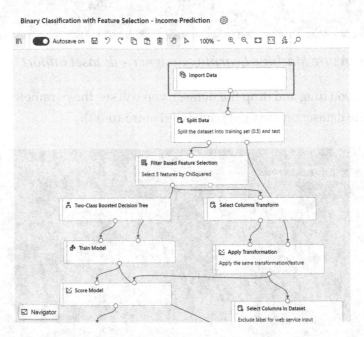

Figure 10-28. *Azure Machine Learning designer – authoring*

2. The designer is a drag-and-drop tool for citizen data scientists, with low-code focus. It is fairly easy to use. You can start by dragging the **Adult Census Income Binary Classification** dataset (as shown in Figure 10-29).

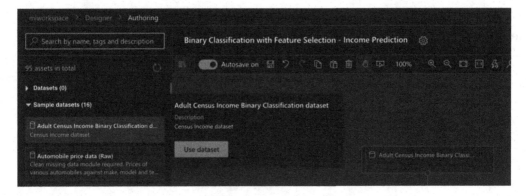

Figure 10-29. *Azure Machine Learning designer – dataset import*

Once you drag and drop the dataset, you will see the parameters for the dataset on the right pane (see Figure 10-30).

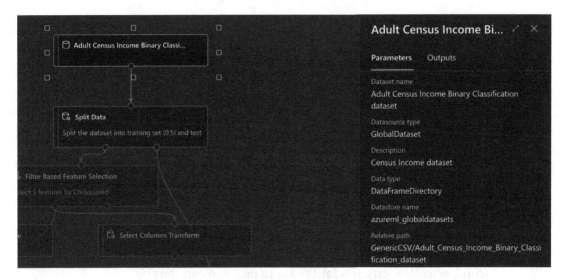

Figure 10-30. *Azure Machine Learning designer – dataset import*

3. The dataset needs to be split into a training dataset and a test dataset. This was accomplished in the previous session by using code. However, here we will use the split data module to accomplish this task. We will perform the randomized split of the dataset. See Figure 10-31.

Figure 10-31. *Azure Machine Learning designer – split data*

4. Then filter the features and define the target column (income), with Pearson correlation as the feature scoring method. The other supported technique is chi squared, as seen in Figure 10-32.

Figure 10-32. *Azure Machine Learning designer – Filter Based Feature Selection*

5.   Now that the data ingestion is completed, a Two-Class Boosted
     Decision Tree module can be used to initialize the classifier, which
     requires a train model module and associated transformation. In
     the left pane, you can see the different types of transformations
     that are supported by the Azure Machine Learning designer (as
     shown in Figure 10-33).

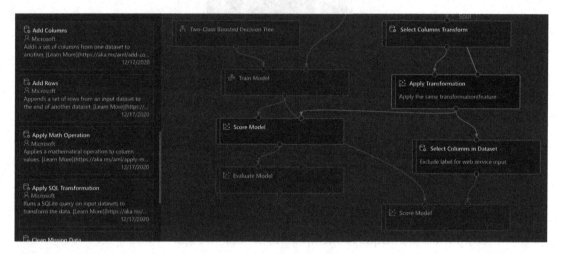

Figure 10-33. *Azure Machine Learning designer – select columns*

Connect the Train Model module with the Score Model, along with the test set. Add
the Evaluate Model module. The final pipeline should look like Figure 10-34.

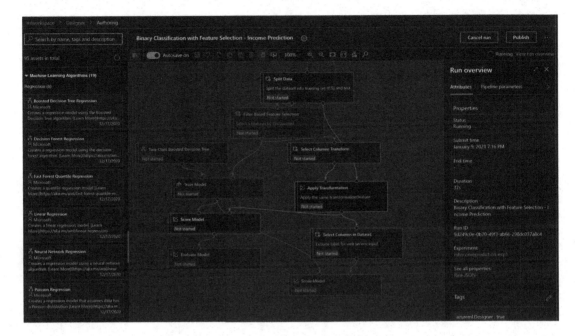

Figure 10-34. *Azure Machine Learning designer – final flow*

On the left pane, you can see the list of algorithms that are available, including boosted decision trees, fast forest regression, linear regression, neural networking regression, and Poisson regression, to name a few. As the run completes, you can now see the run attributes and the score card for model used, in this case, the Two-Class Boosted Decision Tree (see Figure 10-35).

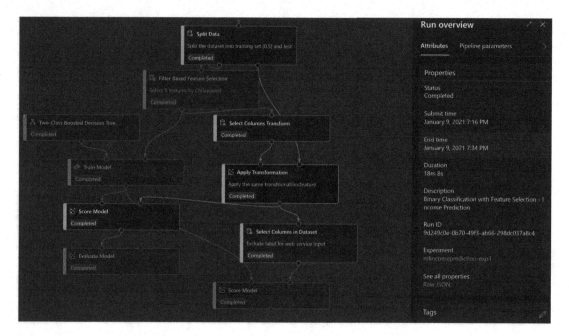

Figure 10-35. *Azure Machine Learning designer – Run overview*

Once the run is completed, in the model runs overview window, you can expand and see the individual tasks within the run (evaluate model, applying transformation, scoring, selecting columns, splitting dataset, and so on) and the time taken for each individual part of the process. Here, we can see that the Two-Class Boosted Decision Tree took the most time. You can also use Azure Machine Learning Visual Studio Code extension; step-by-step instructions are found here.[5] See the Run view in Figure 10-36.

---

[5] Set up Azure Machine Learning Visual Studio Code extension (preview), at `https://docs.microsoft.com/azure/machine-learning/tutorial-setup-vscode-extension`

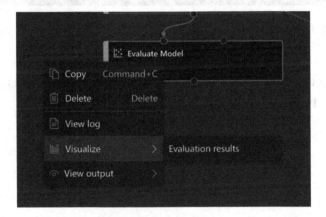

Figure 10-36. *Azure Machine Learning designer – Run view*

Now that the run has been completed, right-click the **Evaluate Model**, and click **Visualize** and then **Evaluation results**, as shown in Figure 10-37.

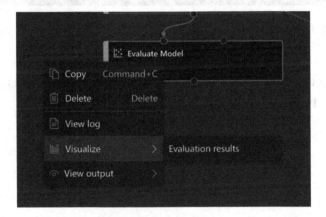

Figure 10-37. *Azure Machine Learning designer – Evaluate Model visualization*

The results show the Score Model result visualization for different attributes and their scored label (such as whether they would have income greater or less than $50,000). See Figure 10-38.

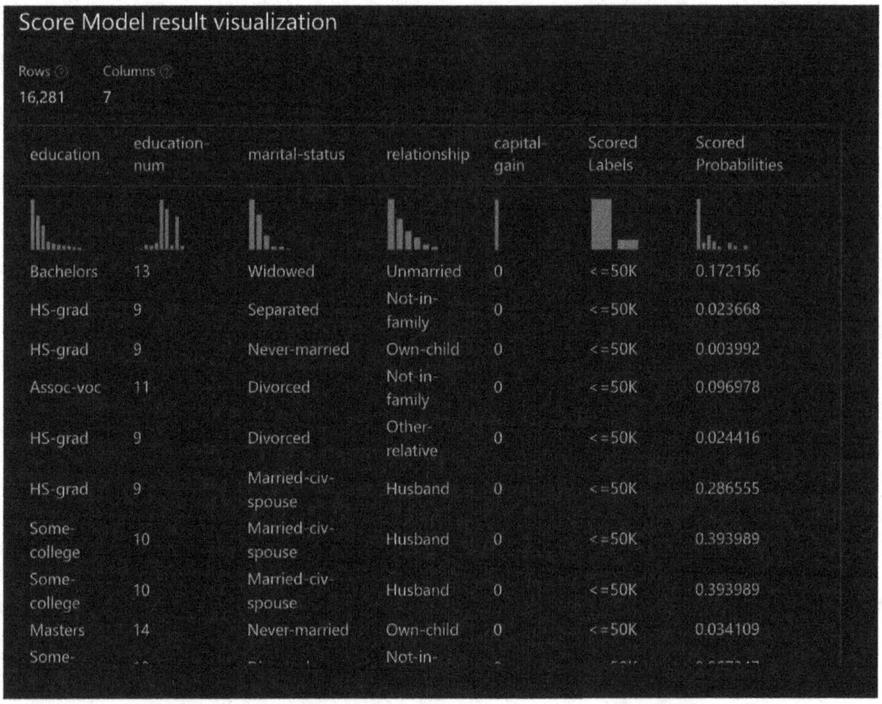

Figure 10-38. *Azure Machine Learning designer – Score Model result visualization*

You can also see the accuracy, precision recall, lift, and confusion matrix for the model (see Figure 10-39). In this case, the model accuracy is ~85%.

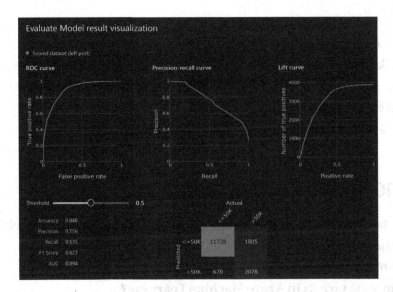

Figure 10-39. *Azure Machine Learning designer – Evaluate Model result visualization*

In this low-code, designer-based, drag-and-drop approach, you can get the results faster with minimal coding. This is a suitable IDE for citizen data scientists, and perspective data scientists and machine learning engineers, who want to experiment with the given dataset, without putting in a lot of effort.

Conclusion and Summary

You have made it to the end of this book. We hope it was a meaningful journey for you, and it was helpful to your goals of learning and exploring the Azure Cognitive Services and Azure Machine Learning landscape.

In this final chapter, you've learned about Azure Machine Learning offerings, beyond Cognitive Services, getting started with Azure Machine Learning, and the machine learning services ecosystem. You have configured, built, deployed, and tested a classification model by using Azure Machine Learning Jupyter notebook. You explored capabilities such as collaborative notebooks, machine learning operations, reinforcement learning, automated ML, data labeling, and autoscaling compute.

Now you can apply the knowledge you have learned into real-world projects. The best way to learn these new, ever-changing, and nuanced technologies is to start working with them. Like AI and machine learning as a discipline, Azure Machine Learning is constantly evolving and adding new capabilities, features, and skills. Today is a good day to start (or to continue learning).

Happy machine learning!

References and Further Reading

Python notebooks with ML and deep learning examples with Azure Machine Learning | Microsoft

https://github.com/Azure/MachineLearningNotebooks

What are compute targets in Azure Machine Learning?

https://docs.microsoft.com/azure/machine-learning/concept-compute-target

Use automated ML in an Azure Machine Learning pipeline in Python

https://docs.microsoft.com/azure/machine-learning/how-to-use-automlstep-in-pipelines

Index

A

Adult income prediction, 344
AIOps, 214
Anaconda navigator environment, 209
Anomaly detector service, 15
    Anaconda navigator, 268
    Anaconda navigator environment, 209
    API, 199, 200
    application starts serving, 266
    batch invocation, 205
    cognitive services, 202
    daily sampling frequency, 213
    demo, 201
    deployment completion, 203
    detect method, 211
    hourly sampling frequency, 212
    initializing prerequisites, 210
    keys and endpoint, 206, 265
    latest point anomaly detection, 268
    notebook, 210
    notebook request, 207
    notebook response, 208
    quick start, 204
    resource access key, 269
    result, 270
    statistical approaches, 199
    Swagger page, 267
    validation completion, 203
    welcome screen, 266
Artificial intelligence (AI), 1, 37, 38, 291

Association for Computational
    Linguistics (ACL), 115
Audacity
    changing format, 140
    changing mix, 141
    changing rate, 140
    file format changes, 141
    format details, 139
    opening, 139
Automated machine
    learning (AutoML), 326
Azure Cloud Shell, 276, 277
Azure cognitive search
    add cognitive skills, 228, 229
    create indexer, 232
    create instance, 225
    create offering screen, 224
    customize target index, 231
    dashboard, 227
    data enrichment
        capabilities, 223
    deployment completed, 226
    enrichments, 229, 230
    import data, 227, 228
    import dataset, 228
    results, 224
    select pricing tier, 225
    testing
        additional documents (hotels), 239
        autocomplete invocation, 238
        create demo app, 234

355

N